# European Conference on Process Safety and Big Data 2020

AA009027

Online
13-14 October 2020

ISBN: 978-1-7138-6095-2

**Printed from e-media with permission by:**

Curran Associates, Inc.
57 Morehouse Lane
Red Hook, NY 12571

**Some format issues inherent in the e-media version may also appear in this print version.**



Printed with permission by Curran Associates, Inc. (2022)

For permission requests, please contact AIChE
at the address below.

AIChE
120 Wall Street, FL 23
New York, NY 10005-4020

Phone: (800) 242-4363
Fax: (203) 775-5177

www.aiche.org

**Additional copies of this publication are available from:**

Curran Associates, Inc.
57 Morehouse Lane
Red Hook, NY 12571 USA
Phone: 845-758-0400
Fax: 845-758-2633
Email: curran@proceedings.com
Web: www.proceedings.com

# TABLE OF CONTENTS

## KEYNOTES

Smart Safety Data ........ 1
*K. Lemos*

Digital Transformation in the Chemical Industry - The Devil's in the Details ........ 11
*L. Colegrove*

Insight Driven Chemical Company: How Data Transforms DSM ........ 21
*P. Dew*

Total Digital Factory ........ 29
*F. Gimenez*

## PANELS

Customer Success Factors in Getting Started on Big Data & Analytics ........ 40
*A. Aldeeb, T. Downes*

Challenges to the Process Industry Operators for Leveraging Big Data for Process Safety ........ 50
*K. Krishna, O. Peeters, G. Nesmith, R. Pawulski*

## SESSION 1: BIG DATA ANALYTIC

Rethinking Process Safety & Risk Management in Operations - The Emergence of ORM Digital Twins ........ 60
*S. Jones*

Open Source Data Mining and Machine Learning for Process Alarms of SCADA Systems ........ 68
*G. Dorgo, T. Ruppert, A. Palazoglu, J. Abonyi*

## SESSION 2: RISK ASSESSMENT & MANAGEMENT

HAZOP+: Combining HAZOP and Operationan Optimisation by Machine Learning ........ 75
*H. Schwarz*

HSE Management in the Age of Artificial Intelligence. A Brightsite Program ........ 87
*J. Van Middelaar, S. Shahmohammadi, E. de Goede, J. Van Den Hurk, G. Hoedemakers*

Knowledge Driven Online FMEA System for Risk Management ........ 95
*N. Ranchet*

## SESSION 3: PREDICTIVE MAINTENANCE

Preventive Maintenance & Machine Learning - Preventive Maintenance by Making use of Machine Learning and Currently Available Instrumentation ........ 101
*B. Groenenboom*

Asset Health Monitoring - Assessing Mechanical Asset Integrity by Means of Permanently Mounted Sensors ........ 110
*C. Buque*

Process Safety: Understanding how AI can Drive Value from Big Data ........ 117
*J. Chappell*

## SESSION 4: ROBOTIC & EMERGING TECHNOLOGIES

Ground Robots can Lead to a Step Change in Safety ........ 124
*D. Drenth, M. Biegl*

Technician of the Future - Taking your Work Management Processes to the Next Level ........ 139
*H. Purohit, K. Harding*

Delek US - Refinery of the Future ........ 151
*L. LaBella, J. Petrusich*

**Author Index**

2020 EUROPEAN CONFERENCE ON PROCESS SAFETY & BIG DATA

October 13–14, 2020

CCPS EPSC

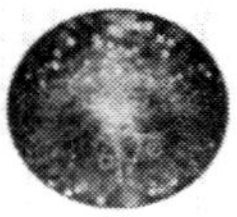

VIRTUAL

# Welcome from CCPS & EPSC!

**Shakeel Kadri, Executive Director & CEO, CCPS**

**Dr. Hans V. Schwarz, CEO, ProsafeX & EPSC Board Member**

2020 EUROPEAN CONFERENCE ON PROCESS SAFETY & BIG DATA

October 13–14, 2020

CCPS EPSC

VIRTUAL

2016 EUROPEAN CONFERENCE ON PROCESS SAFETY AND BIG DATA CCPS EPSC

2017 EUROPEAN CONFERENCE ON PROCESS SAFETY AND BIG DATA CCPS EPSC

2018 EUROPEAN CONFERENCE ON PROCESS SAFETY & BIG DATA CCPS EPSC

2019 EUROPEAN CONFERENCE ON PROCESS SAFETY & BIG DATA CCPS EPSC

**Our journey**

VIRTUAL

# 2020 EUROPEAN CONFERENCE ON PROCESS SAFETY & BIG DATA

October 13 – 14, 2020

CCPS EPSC

The 5th Edition of the European Conference on Process Safety and Big Data will be featuring Augmenting Process Safety Performance through Big Data, Digitalization and Robotics. Data generation & digitization is growing fast and can help to improve the safety of hazardous chemical operations.

SESSION TOPICS

- Keynotes
- Panel Discussions
- Big Data Analytics
- Risk Assessment & Management
- Predictive Maintenance
- Robotics & Emerging Technologies

➢ 130 Registrants, 76 Companies

*Co-organized by the Center for Chemical Process Safety and the European Process Safety Centre*

# Thanks to Planning Committee Members:

- Abdul Aldeeb – Siemens
- Joep Coenen – Versatec
- Tony Downes – Honeywell
- Arthur Groot – DSM
- Pol Hoorelbeke – Total
- Tijs Koerts – EPSC
- Shakeel Kadri – CCPS
- Jing Chen - CCPS
- Daniel Mottaghi – Total
- Richard Pawulski – Air Liquide
- Ricardo Rendall – Dow
- Hans Schwarz – ProsafeX
- Mike Snyder – DEKRA
- Joe Stough – SIFZero
- Willi Meier – DECHEMA / CCPS
- Horst Massong – DECHEMA / CCPS

***AIChE Staff:***

- Evan Flach – Associate Director, Sales

# THANK YOU TO OUR CONFERENCE SPONSORS

## Gold Sponsor

AcuTech
PROCESS RISK MANAGEMENT

## Silver Sponsors

AVEVA GOARC Safety 4.0 sphera™

# Conference Recording and Proceeding will be Available after the Conference

**Conference Recordings:**

- Available for 30 days after the conference on virtual platform
- https://aiche-ecps20.vfairs.com/

**Conference Proceeding:**

- Available on AIChE/CCPS Website
- http://www.aiche.org/ccpseuropeanconference

2020 EUROPEAN CONFERENCE ON PROCESS SAFETY & BIG DATA

October 13 – 14, 2020

CCPS EPSC

VIRTUAL

# Today's keynotes

Katherine A. Lemos, Ph.D.

Lloyd F. Colegrove, Ph.D.

2020 EUROPEAN CONFERENCE ON PROCESS SAFETY & BIG DATA

October 13 – 14, 2020

CCPS EPSC

VIRTUAL

# Thank You and Enjoy the Conference!

THANK YOU TO OUR CONFERENCE SPONSORS

Gold Sponsor

AcuTech
PROCESS RISK MANAGEMENT

Silver Sponsors

AVEVA GOARC Safety 4.0 sphera™

# An Introduction to EPSC

Conference 'Process Safety & Big Data'

EPSC
THE PROCESS SAFETY NETWORK

Hans Schwarz
EPSC Boardmember

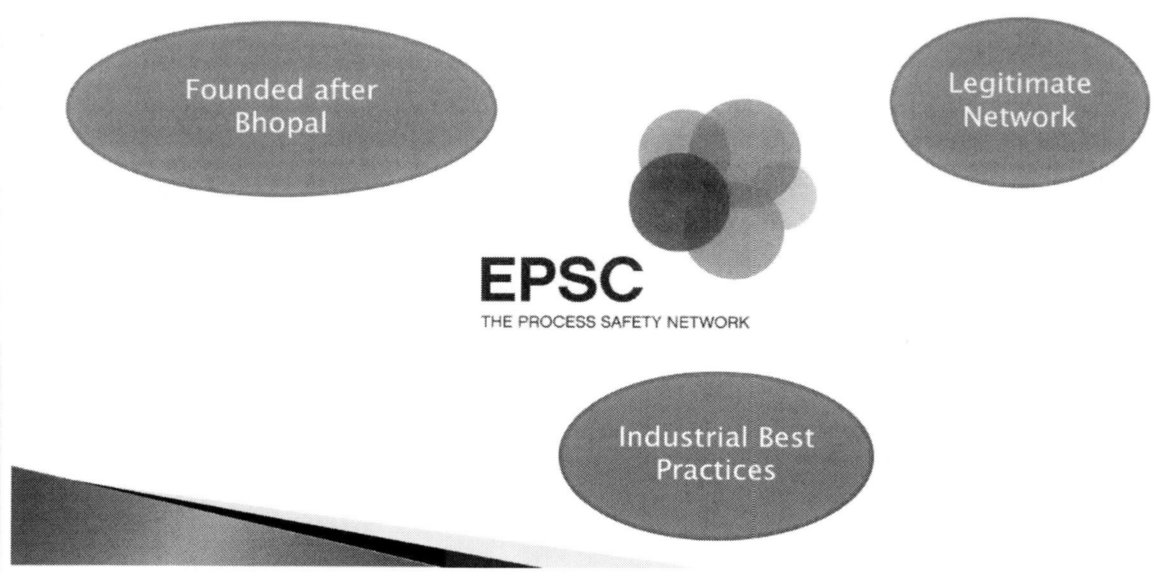
Founded after Bhopal
Legitimate Network
EPSC
THE PROCESS SAFETY NETWORK
Industrial Best Practices

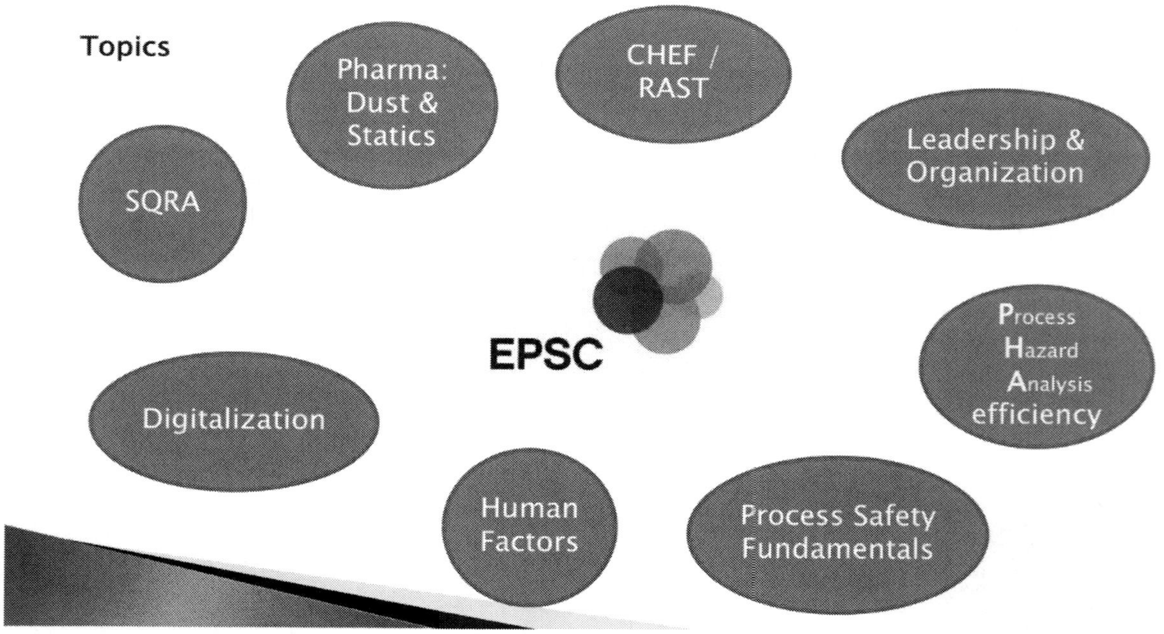
Topics
Pharma: Dust & Statics
CHEF / RAST
Leadership & Organization
SQRA
EPSC
Process Hazard Analysis efficiency
Digitalization
Human Factors
Process Safety Fundamentals

# EPSC ProcessSafetyFundamentals Pictograms

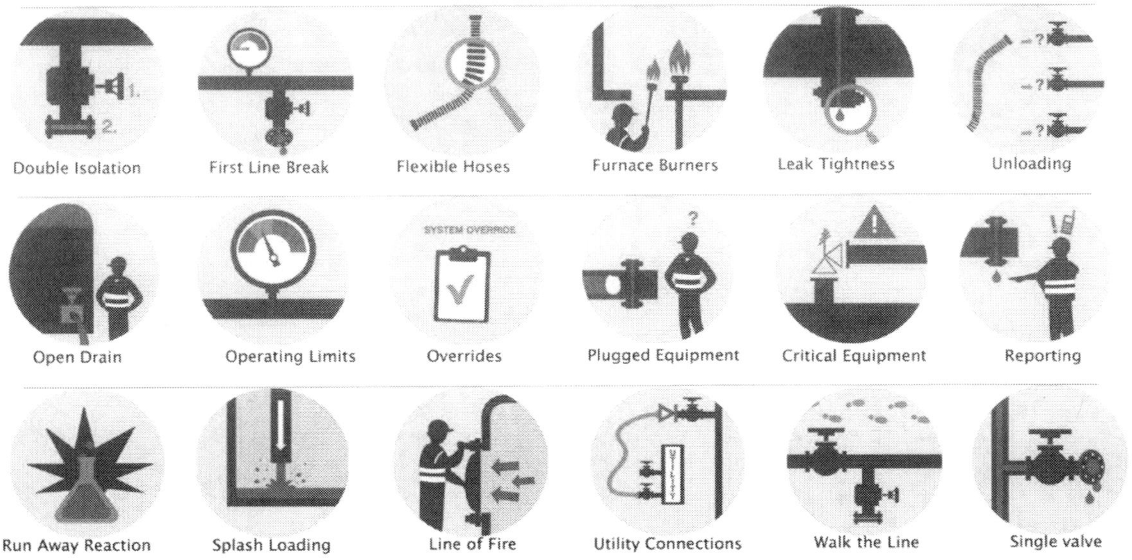


## The Network

- Two day Technical meeting in spring and autumn
- Valuable information (documents) with web access
- Working Groups
- Training material

## Organisation

- Chosen board: DSM, BASF, Covestro, Total, LyondellBasell, GSK, Evonik
- Founding members: Dechema & EFCE
- Operation Manager & Secretary at Dechema
- Full members
- Associate members

EPSC

EPSC

Sharing practical learnings
Antwerp event: Dec 7&8 2021
Contribute with a speaker
Special offer for members
www.SafetyCongress.eu

ProcessNet VCI essenscia ie-net SFGP
BG RCI AIDIC VNCI EFCE VDI
energy institute GESIP cefic



## Learning Sheets

EPSC

www.EPSC.be

### Imploded Storage Tank

EPSC Learning Sheet, June 2019

EPSC

What Happened:
After cleaning with steam a tank was quenched with cold water to shorten the cooling time (not normal practice). The tank vent was not calculated for the vacuum resulting from this rapid quenching operation, resulting in a collapsed tank (without liquid release).

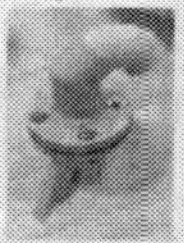

A larger vent size was installed after the incident

Imploded storage tank

Aspects:
- Atmospheric storage tanks are not designed to withstand external pressure on the tank wall. A pressure as low as 20 mbar below atmospheric pressure can cause tank damage
- The force on the wall can be high even at low pressure as the tank wall surface area is very large: Force = Area x delta P
- Consider all scenarios resulting in pressures below atmospheric and make sure vent sizes are adequately designed for these, including abnormal operations
- Make sure vents and safety devices remain operable: fouling bird nests, plastic bags, have all caused tank implosions

Atmospheric tanks are usually not designed for under pressures and can be damaged easily

# EPSC Members

Total, Nouryon, Eastman, LyondellBasell, Dow, Linde, Clariant, MSD, Linde, Avebe, DSM, Evonik, Sasol, AkzoNobel, Huntsman, Eni, Boehringer Ingelheim, Covestro, DuPont, BakerRisk, Corteva agriscience, MOL Group, Bayer, Petrom, TÜV SÜD, gsk GlaxoSmithKline, Solvay, BAM, TÜV Austria, Hovione, sanofi aventis, Merck, Royal HaskoningDHV, TÜV Nord, Borealis, Centrient, Tata Steel, Neste, worldsteel association, Chemours, ExxonMobil, SITECH, Cargill, aramco, BASF, Johnson & Johnson, سابك sabic, OCI Nitrogen, Vynova

# Our keynotes today

Conference 'Process Safety & Big Data'

CDO of DSM

***Peter Dew***
Chief Digital Officer DSM
**Keynote 3: *"How Digitalization will Work for DSM to Get Safer Processes"***

CDO of Total

***Frederic Gimenz***
Chief Digital Officer and VP Digital Factory Project, Total
**Keynote 4:**
***"Digitalization at Total"***

Conference Time Zone: *Central European Summer Time*

| **October 13, 2020** | |
|---|---|
| 2:00 – 2:10 PM | **WELCOME ADDRESS:** ***Shakeel Kadri,*** *Executive Director & CEO, CCPS* |
| 2:10 – 3:00 PM<br>Keynotes Session Chair:<br>Shakeel Kadri | **KEYNOTE 1:** ***Dr. Katherine Lemos,*** *CSB Chair & CEO*<br>***Topic: "Smart Safety Data"*** |
| | **KEYNOTE 2:** ***Lloyd F. Colegrove,*** *The Dow Chemical Company, Retired*<br>***Topic: "Digital Transformation in the Chemical Industry, the Devil's in the Details"*** |
| 3:00 – 3:20 PM<br>Facilitator:<br>Hans Schwarz | **Networking Break & Keynote Q&A** |
| 3:20 – 4:10 PM<br>Panel Session Chair & Moderator:<br>Abdul Aldeeb & Tony Downes | ***PANEL 1: Customer Success Factors in Getting Started on Big Data & Analytics Journey***<br><br>**Panelists:**<br>***Joep Coenen,*** *Versatec Energy, Sr. Consultant*<br>***Dr. Payam Jamshidi,*** *Siemens Energy, Global Lead for Mechanical Integrity*<br>***Joe Stough,*** *SIFZero Advisory Services, Founder*<br>***Eugen Vlaicu,*** *Honeywell Connected Industry, Application Engineering Manager* |
| 4:10 – 4:25 PM | **Break** |
| 4:30 – 5:20 PM | **SESSION 1:** Big Data Analytic |
| Session Chair:<br>Arthur Groot | Rethinking Process Safety & Risk Management in Operation- The Emergence of ORM Digital Twins<br>***Simon Jones,*** *Sphera* |
| | Open Source Data Mining and Machine Learning for the Process Alarms of SCADA System<br>***Gyula Dörgő,*** *University of California, Davis* |
| | Predictive Maintenance Using Artificial Intelligence - Fire & Gas Detectors Maintenance Planning Based on Statistical Analysis and Weather Forecast<br>***Stig Pettersen,*** *Lundin Energy* |
| 5:20 – 5:40 PM<br>Facilitator:<br>Willi Meier & Horst Massong | **Sponsored Networking Events**<br>• Improving Process Safety Performance Evaluation with a Cloud-Based Solution - ***AcuTech Sponsored Networking***<br>• Industry Think Tank - ***GoArc Sponsored Networking*** |
| 5:40 – 6:30 PM | **SESSION 2:** Risk Assessment & Management |
| Session Chair:<br>Daniel Mottaghi | *Acutech Sponsored presentation (5 minutes)* |
| | "HAZOP+" Operational Optimisation & HAZOP as a Synergy<br>***Hans Schwarz,*** *ProSafeEx*<br>***Thomas Froese,*** *Atlantec Systems* |
| | HSE Management in the Age of Artificial Intelligence – Predictive Modeling of Process Safety<br>***Johan van Middelaar,*** *TNO*<br>***Sadegh Shahmohammadi,*** *TNO* |
| | A Knowledge Driven Online FMEA System for Risk Management in Petroleum Refinery<br>***Nicolas Ranchet,*** *TOTAL* |

Conference Time Zone: *Central European Summer Time*

| October 14, 2020 | |
|---|---|
| 2:00 – 2:10 PM | WELCOME ADDRESS: ***Hans Schwarz,*** *ProsafeX* |
| 2:10 – 3:00 PM<br>Keynotes Session Chair:<br>Hans Schwarz | KEYNOTE 3: ***Peter Dew,*** *Chief Digital Officer DSM*<br>***Topic : How Digitalization will Work for DSM to Get Safer Processes*** |
| | KEYNOTE 4: ***Frederic Gimenez,*** *Chief Digital Officer and VP Digital Factory Project, Total*<br>***Topic :Digitalization at Total*** |
| 3:00 – 3:20 PM<br>Facilitator:<br>Joe Stough | **Networking Break & Keynote Q&A** |
| 3:20 – 4:10 PM<br>Panel Session Chair & Moderator:<br>Richard Pawulski | ***PANEL 2: Challenges to the Process Industry Operators for leveraging Big Data for Process Safety***<br><br>**Panelists:**<br>***Olivier Peeters,*** *Total, Vice President, Technology and Projects*<br>***Greg Nesmith,*** *Dow, Senior Process Safety Manager*<br>***Kiran Krishna,*** *Shell Global Solutions, Systems Engineering Digital Lead* |
| 4:10 – 4:25 PM | **Break** |
| 4:30 – 5:20 PM | **SESSION 3:** Predictive Maintenance |
| Session Chair:<br>Mike Snyder | Preventive Maintenance by making use of Machine Learning and Currently Available Instrumentation<br>***Bart Groenenboom,*** *DSM Food Specialties* |
| | Assessing Mechanical Asset Integrity by Means of Permanently Mounted Sensors<br>***Buque Cesar,*** *TÜV SÜD* |
| | Process Safety: Understanding how AI can Drive Value from Big Data<br>***Jim Chappell,*** *Aveva* |
| 5:20 – 5:40 PM<br>Facilitator:<br>Tony Downes & Horst Massong | **POSTER SESSION** |
| 5:40 – 6:30 PM | **SESSION 4:** Robotic & Emerging Technologies |
| Session Chair & Conference Closing Remarks:<br>Tijs Koerts | Ground Robots can Lead to a Step Change in Safety<br>***Duco Drenth,*** *Dietsmann* |
| | Technician of the Future: How to take your work management processes to the next level<br>***Heena Purohit,*** *IBM*<br>***Kyle Harding,*** *IBM* |
| | Delek U.S. Vision for Refinery of the Future<br>***Louis LaBella,*** *Delek*<br>***Jim Petrusich,*** *GoArc* |

**Remarks prepared**
**for**
**Katherine Lemos, PhD.**
**Chairman and CEO, U.S. Chemical Safety and Hazardous Investigation Board**
**Before the**
**CCPS 2020 European Conference on Process Safety and Big Data**

**October 13, 2020**

I appreciate the invitation to speak to you today at the CCPS 2020 European Conference on Process Safety and Big Data.

It is an honor to be asked to deliver the keynote lecture, and to represent the U.S. Chemical Safety and Hazard Investigation Board, the CSB, and our mission to "Drive Chemical Safety Change."

I'll start today by providing a brief introduction to:

- the scope and mandate of our agency,
- my professional background and what I bring to my new role as the Chairman, and
- my goals and priorities for this agency.

The theme of my talk today is "Smart Safety Data."

For me, this translates to using data to drive mission actions and efforts, and to do it smartly.

I'll cover a few different aspects of safety data, from

- operator information, to
- safety management systems, to
- a Big Data perspective of safety for the chemical industry.

I'll discuss:

- why this is important, and
- how we can work together to make this vision a reality.

***Agency Introduction***

For those of you not familiar with our agency, let me succinctly define what we are about.

The CSB is charged with investigating inadvertent chemical releases at fixed facilities throughout the United States.

These releases have significant impact on the safety of humans and the environment.

Impact is defined in terms of fatality, serious injuries, or substantive property damage. It also includes significant risk to the community, which would involve displacement or sheltering in place.

Our investigative teams are tasked with:

- reporting the facts, conditions and circumstances of the event;
- determining the probable cause or causes;
- and from this, issuing safety recommendations for action to a range of stakeholders across the community.

Our investigations examine and evaluate a wide range of aspects, to include

- equipment and system design;
- regulations, industry standards and guidance;
- training, operations and procedures; and
- human and organizational factors.

Board Members are responsible for ensuring three things:

- the integrity of our mission products,
- approving staff findings and recommendations, and
- advocating on behalf of the agency's priorities.

However, there is one purpose alone for CSB's activity: To reduce the likelihood of recurrence and consequence of future chemical incidents or accidents.

Let me say that again: To reduce the likelihood of recurrence and consequence of future chemical incidents and accidents.

***Personal Introduction***

I was honored to be nominated by President Trump for two separate positions, as Board Member and as Chairman of the CSB. I was confirmed by the Senate for both of these roles in March.

Prior to joining the CSB, my work has primarily been in the aviation sector –

- within industry at Northrop Grumman,
- as a regulator at the Federal Aviation Administration,
- as an independent investigator at the National Transportation Safety Board, and
- as a researcher at NASA and in academia.

The experience I gained across the stakeholder landscape has provided me with a deep understanding of the following: how parties must work cooperatively together to enable safety improvements.

The amazing thread between all these roles.

It's that success was always dependent on obtaining or deriving safety data to drive decision-making.

That decision-making can be for:

- making recommendations for investments in new research,
- for developing new federal regulations or guidance materials, or
- for enhanced training, or a re-design of systems or process.

All these decisions are dependent on safety data, whether the CEO or the front-line operator is making the decision.

Even when the desired outcome is efficiency, safety must also be considered.

I started my term as Chairman at the end of April – during the early days of the COVID-19 pandemic in the United States. Shortly after I began, I became the only member of the CSB's five-member board.

To say these factors haven't presented challenges would be disingenuous. But as Chairman, my unequivocal goal is to ensure that the core mission work of the agency continues unabated.

I am proud to say that we are working diligently on our investigations and safety products.

***Goals and Priorities***

Our top priorities are the following:

1. To Focus on the Mission: Continue delivering high quality safety product to the community.
2. To Drive efficiency of operations within the agency – includes staffing; and
3. To Leverage and strengthen stakeholder relationships.

I'd like to highlight a recent accomplishment that demonstrates our success in working with stakeholders to drive chemical safety change.

In June we closed a recommendation to CCPS that came out of our investigation into the August 2017, fire at the Arkema Chemical Plant in Crosby, Texas.

Let me share this video with you.

***INSERT/PLAY VIDEO: ARKEMA AND CCPS – 4:53***

Once again, we commend CCPS for their significant contribution to our community with this quality product. This is *exactly* the kind of positive safety change that can happen when the CSB and stakeholders work together cooperatively.

So, let's talk now about safety data!

***Smart Safety Data: Operator Information***

I'll first talk about Operator Information.

This is information that front-line operators use to make decisions, and, when successfully delivered, ensures safety of operations.

Front-Line Operators execute on the mission as defined by the leadership.

Most people think of Operator Information as visual and aural information presented on computer displays.

They also think about how the systems are designed to communicate and interact with the operator.

But it extends to other types of operator information, such as:

- The company and manufacturer policies and procedures,
- How well operators are trained to use the information and the guidance at their disposal, and
- Cultural norms and pressures, whether verbal or non-verbal, actual or perceived.

Collecting this level of detail and range of operational data is foundational to any Big Data analysis of safety risks and trends.

This drives leadership decisions.

And when it matters most is in time-limited situations, where there is no clear answer, and the consequences are high.

I'll use an example from the 2018 Pryor Trust Gas Well Blowout and Fire investigation.

As crew members removed drill pipe from the well to change a drill bit, the well became underbalanced. The faulty alarm system design played a role in operators not realizing a significant amount of gas was entering the well.

Let's look at an excerpt of our safety video on this incident.

***INSERT ANIMATION FROM PRYOR TRUST VIDEO***

As shown in the video, we discovered two things:

- First, on the evening before the blowout, the night shift driller turned off the entire alarm system.
- Second, the day shift driller briefly turned on the alarm system, but then quickly turned it off again on the day of the event.

Other than that short period, the alarm system was off for **14 hours** prior to the blowout.

Although we weren't able to conclusively determine why the drillers turned the alarm system off, we do know that the alarms were perceived as a nuisance or annoyance.

That's a common reason for operators disabling alerts.

Our investigation found an interesting conclusion… had the alarm system been engaged, most alarms that would have activated during these 14 hours would likely have been irrelevant in detecting the actual well problems they were facing.

Now, if the alarm system had been activated, several critical alarms, such as the Gain/Loss alarm, would have alerted operators to a gas influx in the mud pits, and could have prevented the incident.

From my work in aviation, I am deeply aware of issues stemming from the design of alerts and operator information. Unfortunately, I am not surprised to see it occur within the chemical industry.

Particularly for time-limited and safety-critical decisions, humans need the right information, distilled, and easily accessible. It is imperative that operators are receiving relevant, actionable data about the chemical processes they are operating, without informational distractions.

As a result of what we saw at Pryor Trust, the CSB made recommendations to two manufacturers of drilling alarm systems.

We asked them to design the user interface to allow for easy navigation between "state-based" alarm operations.

For example, on a drilling rig, the alerting system could be designed to activate based on the operational phase, such as drilling, tripping, and circulating. Alerts should be activated and prioritized only if they are critical and relevant to the operation at hand.

Our investigation discovered that other types of operator data also played a role in this incident.

These include the lack of detail in the procedures provided to the operators, as well as a company policy on alerts.

I highlight Operator Information, because capturing this type of detail is essential to the integrity of safety data at scale, when analyzing your operations or monitoring your mitigations.

It is critical to note, the subject matter expert on the front line is the only one that can provide you with the context needed to bring meaning to Big Data.

In addition, this person the only one who can provide the level of detail that translates to successful executive decision-making.

Operator information is a vital link to preventing chemical disasters.

### *Smart Safety Data: Safety Management Systems*

Our investigation at Pryor Trust also revealed a challenge that I am learning exists across the chemical industry, which brings me to the second type of Safety Data: Safety Management Systems.

Gaps in Process Safety Management regulations continue to allow the conditions for incidents to occur.

What does this mean? Without a formal process safety system or safety management system:

- Hazards are not identified, ranked and assessed;
- Mitigations are not proposed or analyzed, much less implemented; and
- The impact of these hazards and mitigations are not monitored.

Data from this type of informational feedback loop is not only vital to maintain safe operations, but is essential to any Big Data safety analysis at the level of the company or across the industry.

During our Pryor Trust investigation, we learned there are minimal federal regulations governing the safety of onshore oil and gas drilling operations. As a result, Pryor Trust was not required to perform a hazard analysis.

With no federal safety standards, onshore drilling companies are governed by a "patchwork" of individual state laws.

State laws, in this case, didn't cover the gap. In Oklahoma, where the incident occurred, the laws aren't safety-focused, but instead stress maximizing production and eliminating the loss of hydrocarbons.

Therefore, we made recommendations to OSHA to either

- apply the PSM standard to onshore drilling of oil and gas wells, or
- to develop a new standard specific to onshore drilling.

That recommendation is still open.

This is just one example of the need to reform process safety regulations.

Process safety practices provide the data to leadership required to help them make the right decisions to avoid mishaps, especially for known hazards.

***Smart Safety Data: Chemistry Industry Big Data***

I've mentioned the need for data to drive mission-oriented leadership, from operator information, to safety management systems.

Now I'll address:

- a Big Data perspective of safety for the chemical industry,
- why this is important, and
- how we can work together to make this vision a reality.

I'll start with our Critical Drivers List, then move on to other agency initiatives that will contribute to a Big Data capability for chemical safety.

*Drivers of Critical Safety Change*

The Critical Drivers List includes safety improvements approved and championed by the CSB for intensive advocacy and heightened awareness.

The selection criteria are data-driven and include consideration of:

- The nature of the risk and estimated extent of exposure to workers, offsite populations, and the environment;
- Previous loss and potential for future loss of life, injury, environmental damage, or damage to property;
- Very high risks disproportionately affecting discrete but highly identifiable groups of individuals; and
- Strong concerns of specific sectors of a community or stakeholders.

The recommendation I just mentioned from the Pryor Trust investigation is just one of 38 total recommendations the CSB has issued to drive modernization of process safety regulations.

Twenty of those recommendations have been successfully closed, but 18 are still open.

Because the CSB considers these recommendations to be of the utmost importance, "Process Safety Management Modernization" is one of five "Drivers of Critical Safety Change."

Our Critical Drivers List includes four other areas:

- Combustible Dust Safety
- Emergency Planning and Response
- Preventive Maintenance
- Safe Hot Work Practices

As Chairman I am committed to not only championing these issues, but to driving and creating an even more robust data-driven approach to selecting and evaluating the progress of our Critical Drivers.

As a mission-oriented agency, we are compelled to take every measure to ensure that our efforts are applied to changes that will have the greatest safety impact for the chemical industry.

There are several data related CSB initiatives we believe will help us to better achieve our mission and, at the same time, to further serve the community.

*CSB Big Data Initiatives*

The first initiative focuses on data received from incident reports.

Currently, data from initial reports are housed in an incident database. From here, we make the determination to whether to deploy. We will be investing in the further organization and development of this database to better exploit the information.

For example, to:

- search for trends for hazards already identified, as well as
- hazards which have yet to be identified,
- to create some baseline rates of occurrence, and to
- work towards the development of or alignment with a common taxonomy for causal factors and outcomes.

I'm also enthusiastic about the additional data we'll receive from the community, prompted by our new Incident Reporting Rule. This rule was long overdue.

The rule requires prompt reports to the CSB from facilities that experience an accidental release of a regulated substance or extremely hazardous substance that results in a death, serious injury or substantial property damage.

Our website now has an area for companies to send their incident reports directly to the agency versus going through the National Response Center.

Having a more complete account of incidents will do two things:

- It will better inform our deployments, and then
- It will strengthen our understanding of safety hazards and occurrence rates.

It will do this across the broad domains within the chemical industry.

Importantly, additional insight and tools will help all of us to prevent future chemical incidents.

Properly collecting and using incident data is an amazing opportunity for the CSB. I look forward to further conversations with stakeholders about both the reporting rule and its implementation.

Another initiative that we've committed to developing is a Docket Management System.

As I mentioned before, I previously worked at the National Transportation Safety Board, upon which the CSB was modeled. The NTSB provides detailed information from its investigations to the public through its online Docket Management System.

Information provided by the NTSB in a typical accident docket includes interview transcripts, technical reports, surveillance videos, and other data collected during an accident investigation.

The value to our developing this type of information system at the CSB would be several-fold:

- It will greatly increase our transparency to the public,
- It will provide another medium to amplify our reach, and make the wealth of safety information the CSB already possesses widely available,
- It will allow us, and you as stakeholders, to mine the data for new insights, hazards, and trends and
- It will lend further opportunity to prevent future chemical safety incidents.

By subjecting our investigation and incident data to a common taxonomy of causal factors, conditions and outcomes – can you imagine the potential we will have to better serve the American Public?

This truly would be a data-driven approach to ensure that we make informed decisions regarding our priorities, and to execute with more clarity and confidence on our mission.

As you can see, I'm quite passionate about the topic of data, and how, when collected, organized, managed and analyzed properly, it can be extremely useful in the decision-making process.

***Closing Section***

In closing, my number one priority and duty as Chairman is to execute and fulfill the mission of the CSB.

I can assure you, we will continue to put forth high-quality safety findings and recommendations that serve to prevent future accidents.

To hold ourselves accountable to our mission, we must be data-driven, as this drives leadership decisions and priorities, which flow throughout the organization.

I have laid out the importance of collecting and attending to operator information, and the value that this brings to the context of any Big Data effort.

I have also highlighted the critical role that safety management systems practices play in developing and maintaining safe operations.  A live feedback loop of this information is essential to any Big Data analysis at the level of the company, or across the industry.

For these reasons, we advocate for the modernization of federal process safety regulations to include hazards and gaps we are already aware of.

And finally, I have presented the CSB's vision to drive and create a more robust data-driven approach to meet our mission, and simultaneously extend our reach and transparency. We will need your support.

Together, we can significantly advance the safety of the chemical industry.

Thank you – I am happy to answer your questions.

##

# DIGITAL TRANSFORMATION IN THE CHEMICAL INDUSTRY
## THE DEVIL'S IN THE DETAILS

LLOYD COLEGROVE, PH.D. INDEPENDENT CONSULTANT AND DATA SCIENCE EVANGELIST (RECENTLY RETIRED FROM DOW)

## AGENDA

- Culture Change and Technical Context.
- Briefly review two commentaries on AI in the Chemical Industry:
  - *Artificial Intelligence in the Chemical Industry – why my industry puzzles over our vendor's struggles* - Journal of Advanced Manufacturing and Processing (Vol 2, Issue 3, July 2020, L.F. Colegrove)
  - *The Challenges of AI* (in the chemical industry) CEP July 2020. (L.F. Colegrove)
- Some practical advice:
  - AI and Contextualization – The "chicken and the egg".
  - Before you do anything ... GET THE NOISE OUT OF YOUR PROCESS DATA!
- Conclusion.

## BRIEF INTRODUCTION.

Lloyd F. Colegrove
https://www.linkedin.com/in/lloydcolegrove/

Ph.D. in Chemical Physics (TAMU)
29+ years in the chemical industry.
22 years in manufacturing.
Led the development of several data science groups in Dow, along with a fundamental problem-solving capability (both) to attack chronic or mysterious problems in our plants.
Married (also Ph.D, R&D Fellow @ Dow) with two daughters - one a Sr. in Physics, the other a graduate student in Psychology/Neuroscience.

When I work(ed)...

When I play...

## IF YOU TAKE NOTHING ELSE AWAY FROM THIS VIRTUAL PRESENTATION:

**Context** - if you cannot use new technology in proper context to the opportunity, you will fail. In the example before you, failure to be able to contextualize your data properly, means no amount of cool tools or software will help you.

**Culture** - if you do not prepare your workforce, your transformation will be extremely slow, and very, very messy. Honestly, you may not ever get where you want to go.

WHEN I SAY:

ARTIFICIAL INTELLIGENCE IN THE CHEMICAL INDUSTRY.

YOU MIGHT THINK:

DATA PROCESS IMPROVED OPERATIONS! AUTOMATION

CONTROL AUTONOMOUS CONTROL THE PLANT WILL RUN ITSELF! I CAN GO HOME AND (FILL IN THE BLANK HERE*)!

*SPEND MORE TIME WITH FAMILY.
GO FISH
BECOME AN ACTIVIST FOR SOCIAL CHANGE
SLEEP IN
ETC.

**NO SO FAST ...**

In our rush to apply the latest and greatest in analytics we have failed to

- bring our workforce along
- strengthen our workforce's understanding of our processes via the fundamentals (engineering or chemistry)
- challenge our current data paradigm
- clean up our process data
- Improve our integration of various data streams into one easily useable and interchangeable data platform.

This just to name a few......

## ARTIFICIAL INTELLIGENCE IN THE CHEMICAL INDUSTRY.

JOURNAL OF ADVANCED MANUFACTURING AND PROCESSING — AIChE

### MAIN POINTS:

- AI companies lack domain knowledge.
- They think they can solve all your problems if you'll just load all your data into their clouds, but when they don't understand your problem they grow frustrated.
- The AI sales pitch rings hollow because the data sets our industry possesses aren't complete enough to hand over to AI alone, even if other seemingly insurmountable problems were addressed.

## AI IN THE CHEMICAL INDUSTRY – A FOLLOW UP ARTICLE ON CULTURE

AIChE

### MAIN POINTS

The op/ed is about the existing culture in our industry and how it must change.

- Gaining workforce acceptance must be a goal of top management.
- Data Acuity (or Data Literacy) must improve across all members of the operational team.
  - Undergraduates must make a better effort to learn the "art of the possible" in analytics.
  - Transforming existing culture while new hires come in is required.
- Know what you are trying to accomplish first!
  - Don't pick the tool first, pick the problem.
  - Find suitable projects for data science, but don't forget to write and advertise success stories in the aftermath of a successful hunt! Then market that success within your company.

## AI IN THE CHEMICAL INDUSTRY – A FOLLOW UP ARTICLE ON CULTURE

AIChE

### MAIN POINTS

The op/ed is about the existing culture in our industry and how it must change.

- Gaining workforce acceptance must be a goal of top management.
- Data Acuity (or Data Literacy) must improve across all members of the operational team.
  - Undergraduates must make a better effort to learn the "art of the possible" in analytics.
  - Transforming existing culture while new hires come in is required.
- Know what you are trying to accomplish first!
  - Don't pick the tool first, pick the problem.
  - Find suitable projects for data science, but don't forget to write and advertise success stories in the aftermath of a successful hunt! Then market that success within your company.

## CASE IN POINT – WHY DIGITAL TRANSFORMATION IS MORE ABOUT PEOPLE THAN TOOLS.

### READY?

## WHICH CAME FIRST, THE CHICKEN OR THE EGG?

Process engineering
Process Control
Plant Design
Chemistry
Kinetics
Physics.

Artificial Intelligence
Data Science
Statistics
Excel
JMP
Etc.

## FORTUNATELY FOR US, THEY ARRIVED AT THE SAME TIME.

**CONTEXT**

**DATA**

## DATA SCIENCE DOES NOT WORK WITHOUT CONTEXT!

- It's easier to turn an engineer or a chemist into a data scientist than it is to turn a data scientist into an engineer or a chemist.
- Data science is worthless if you cannot tie the signals extracted from the data, to the actual physical realities of the plant. (Design, Chemistry, Kinetics, Physics, etc.)
- If you fail to bring your workforce along on this journey, you **will fail** to successfully scale the new tools and approaches that are sure to come.

## AN EXAMPLE SUCCESS STORY

NO SO FAST ...
In our rush to apply the latest and greatest in analytics we have failed to

- bring our workforce along
- broadly understand our processes via the fundamentals (engineering or chemistry)
- challenge our current data paradigm
- **clean up our process data**
- improve our integration of various data streams into one easily useable and interchangeable data platform.

## HOW DO YOU CLEAN UP YOUR DATA WHILE RUNNING YOUR PLANT?

**COLLECT IT, DUMP IT INTO JMP OR EXCEL,
TAKE DAYS TO WEEKS CLEAN YOUR DATA
THEN LOOK AT IT AFTER THE FACT?**

**OR**

**CLEAN YOUR DATA BY RUNNING YOUR
OPERATION BETTER SO THAT YOU DON'T HAVE TO
TAKE A LOOK AT IT AFTER AN "OOPS".**

## CLEAN UP YOUR PLANT DATA WHILE RUNNING THE PLANT.

**WHERE OPERATIONS STARTED**

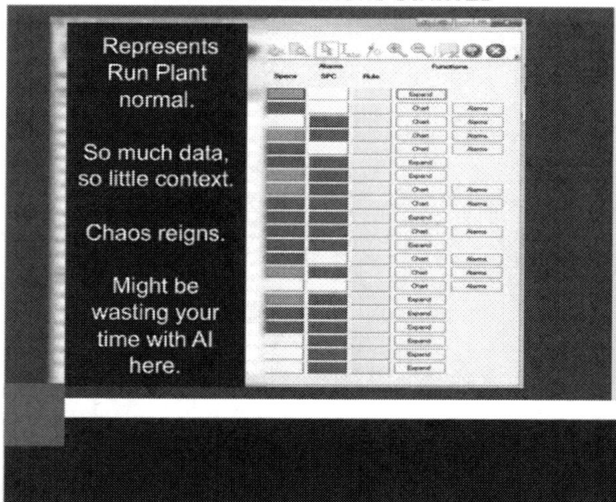


**DIGITAL TRANSFORMATION**

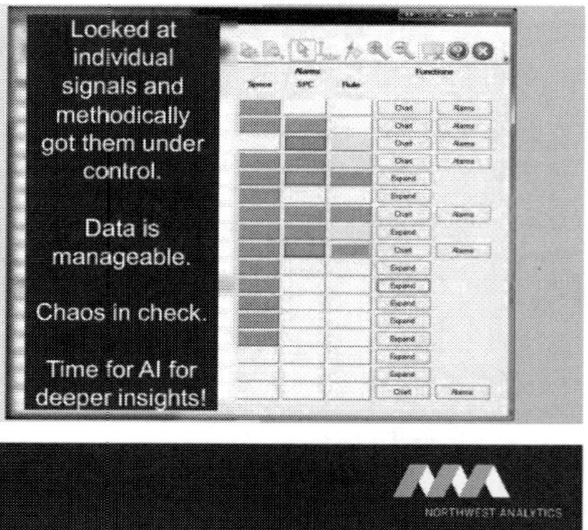


NORTHWEST ANALYTICS

## THE ANALYTICS SHOULD ENABLE PEOPLE, NOT THE OTHER WAY AROUND!

- Force the operations staff to reconcile out-of-control or out-of-tolerance running conditions.
- Shine the bright light of awareness on historic questionable or bad performance of certain data streams.
- Most of the game was getting the data in a form that that operations could digest easily.
- All streams required logging of interactions and interventions to preserve information for future generations.
- We need to do everything possible to provide the right tools for our workforce and set the right expectations for the use of the data that should drive their decision making.
- Prepared a way for the future application of Artificial Intelligence.
- Raised the data equity of the workforce in the process.

## ENTERPRISE DATA VIEW Global view to shop floor in one system

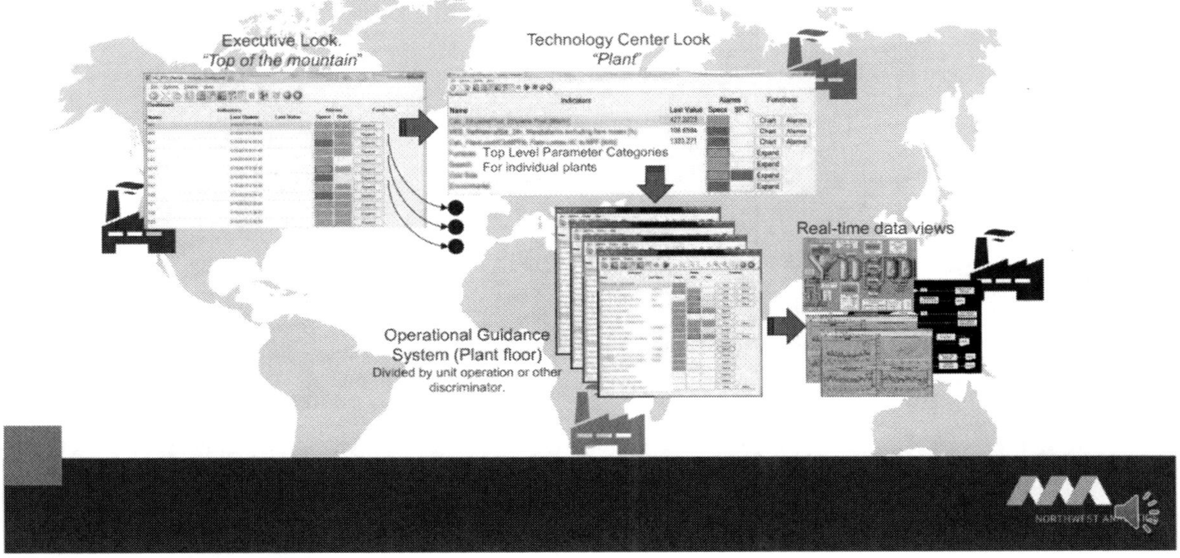

## DIGITAL TRANSFORMATION IS TOUGH.

- It involves people, that must be brought along on the journey, not abandoned on the side of the road.
- Proper target acquisition.
- Agile follow-up and reporting.
- Flexibility to engage the right tool to the right problem.
- Keeping the longer term view, but being grateful for short-term gain.

## THANK YOU!

DIGITAL TRANSFORMATION, THE DEVIL'S IN THE DETAILS.

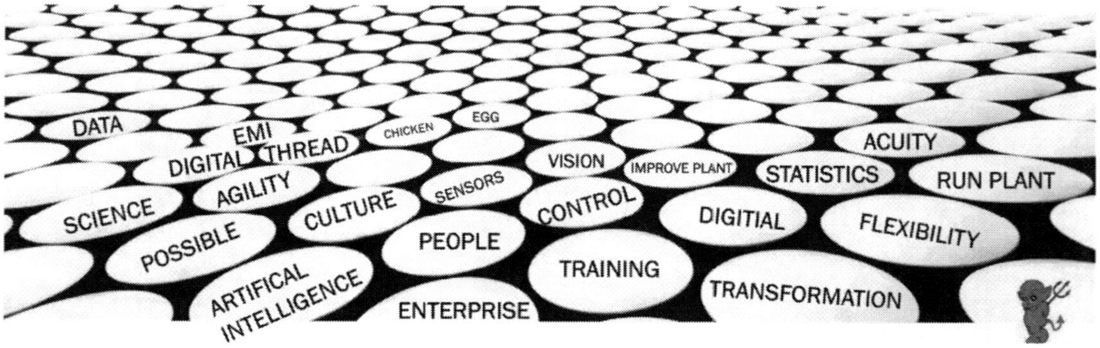

Insight Driven Chemical Company
How data Transforms DSM
Peter Dew, Chief Digital Officer
2020 European Conference on Process Safety & Big Data
October 14, 2020
DSM
BRIGHT SCIENCE. BRIGHTER LIVING.

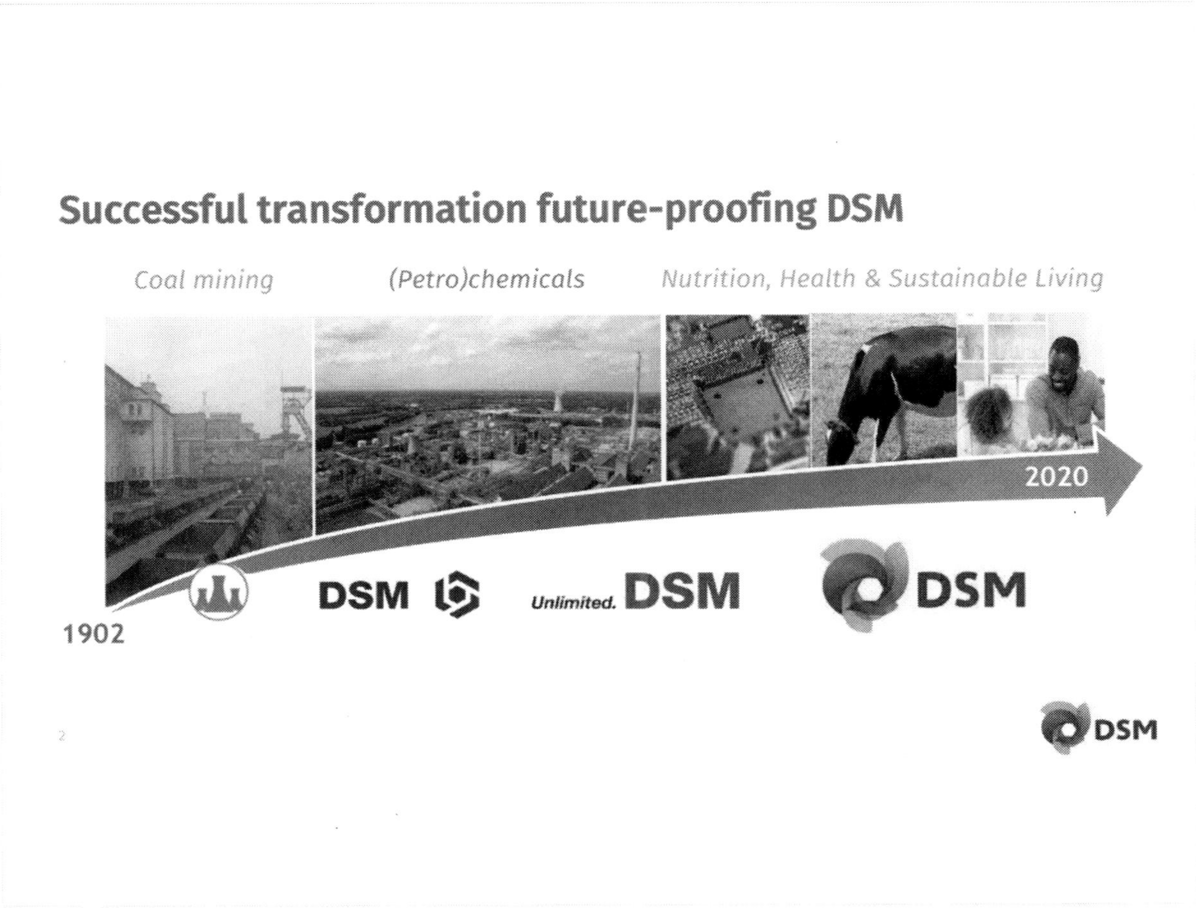
Successful transformation future-proofing DSM
Coal mining
(Petro)chemicals
Nutrition, Health & Sustainable Living
2020
1902
DSM
Unlimited. DSM
DSM
2
DSM

## Driving Digital & Advanced Analytics across DSM

*Looking back and beyond*

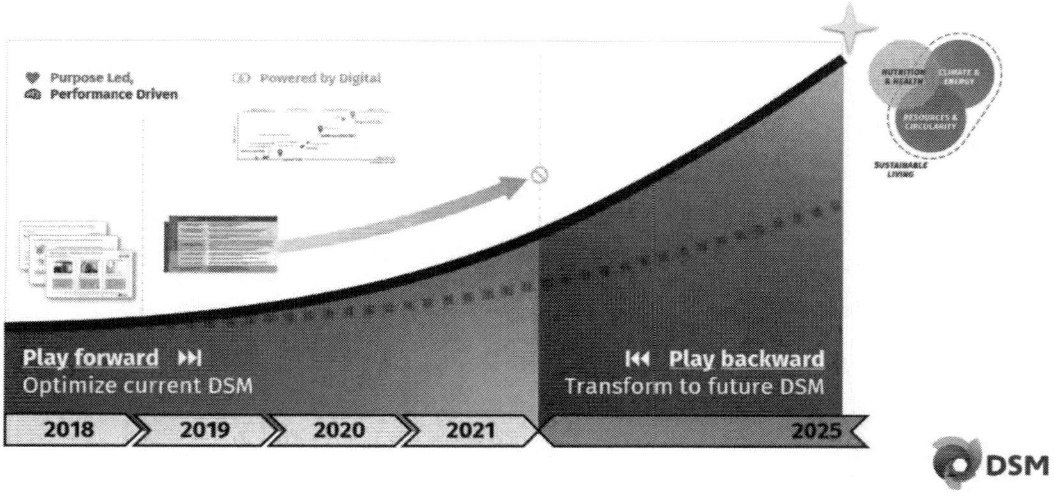


4

## Digital shifts identified across the value chain

| | |
|---|---|
| Automation at scale | **Lead the charge in automation** - eliminate all routine manual processes |
| Data analytics at scale | **Insights driven decision making at scale** |
| Truly agile in business | **Become truly agile** - move from a traditional, waterfall organization to an insight driven organization |
| Future workforce | **Prepare for a future workforce** that will be working cross-functionally (agile) & possess entirely different expertise, habits and routines |
| Strategic partnership building | **Build the right partnerships at scale** and go beyond current business models |



DSM

## Digitalization at DSM Operations

*Simplify and standardize*

**Factsheet**

- > 11.000 people working in DSM Operations
- > 100 different sites worldwide +/- 500 factories
- Data transparency and insights will drive performance
- Facilitates compliance/critical work process
- Operational excellence with advanced analytics

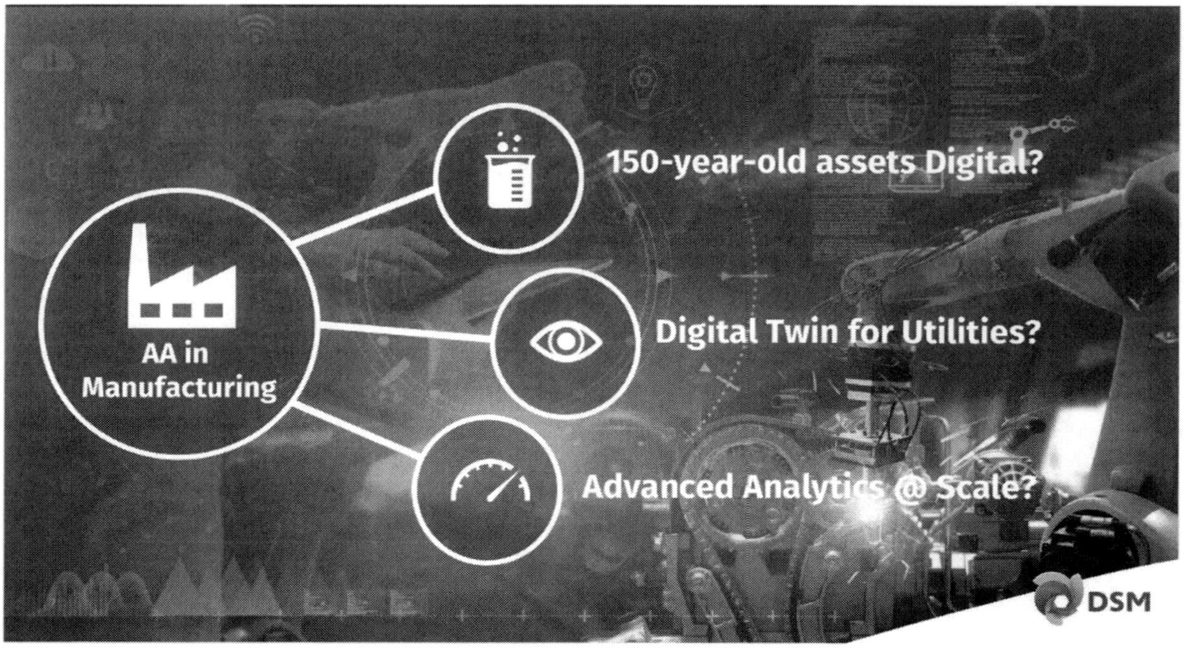
AA in
Manufacturing
150-year-old assets Digital?
Digital Twin for Utilities?
Advanced Analytics @ Scale?
DSM

Bringing a 150-year old factory into the digital age
DSM

## Operational efficiency using digital & data science

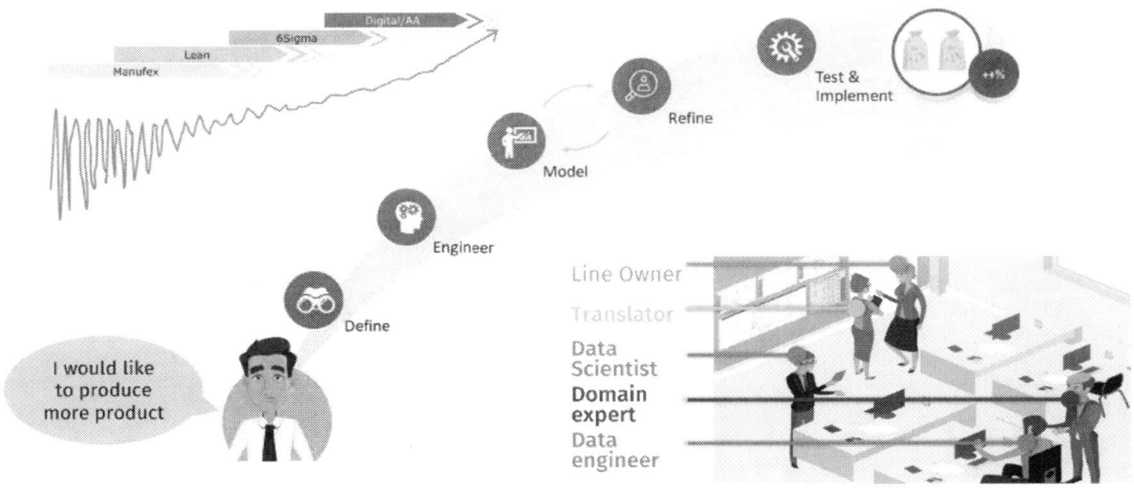


## Energy efficiency enabled by Data Science

*Digital Twin for Utilities*

At DSM data scientists have developed and implemented a Utility model that build a Digital Twin of the Utility site which enables:

- Better management of electricity
- Better management of steam
- Operators (utility managers) to work more efficiently
- Continuous process improvements

Resulting in significant savings a year

Sustainability – smaller carbon footprint

Project is being scaled to multiple sites of DSM

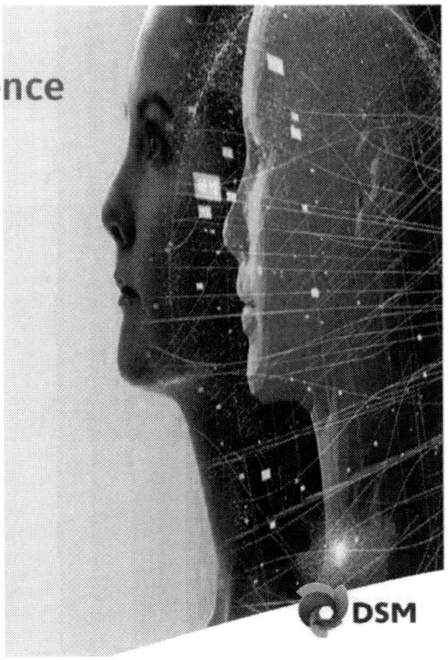

## Digital & Advanced Analytics @ Scale

*Continuous improvement enabled by data*

**Multiple M&T sites** with successfully implemented Advanced Analytics projects, as well as digital and analytics embedded in the DICI methodology. We currently have a roadmap on 20+ cases to scale.

**Yield & Throughput cases**

How can we use Digital & AA to optimize the Yield or throughput of our plants

**Digital Way of Working (DWoW)**

How can we make us of Digital to improve the way of working of our people in operations

DSM

## OE powered by Digital Way of Working & Advanced Analytics

*Leading to continuous improvements in Process Safety*

**Automated workflows**

leads to less manual interference – hence less errors / risks

**Machine Learning Models**

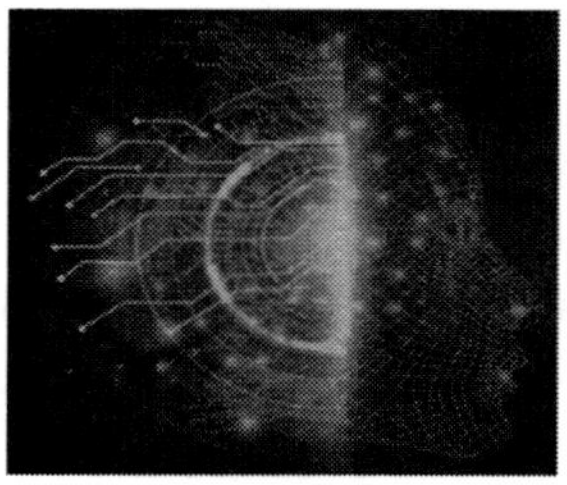

predict scenarios before they happen - hence less errors / risks

**Digital Twin**

enables DSM to continuously optimize performance without being on site – hence less errors / risks

12

DSM

## Next steps

| | |
|---|---|
| Automation at scale | **Lead the charge in automation** - eliminate all routine manual processes |
| Data analytics at scale | **Insights driven decision making at scale** |
| Truly agile in business | **Become truly agile** - move from a traditional, waterfall organization – to an insight driven organization |
| Future workforce | **Prepare for a future workforce** that will be working cross-functionally (agile) & possess entirely different expertise, habits and routines |
| Strategic partnership building | **Build the right partnerships at scale** and go beyond current business models |

13

DSM

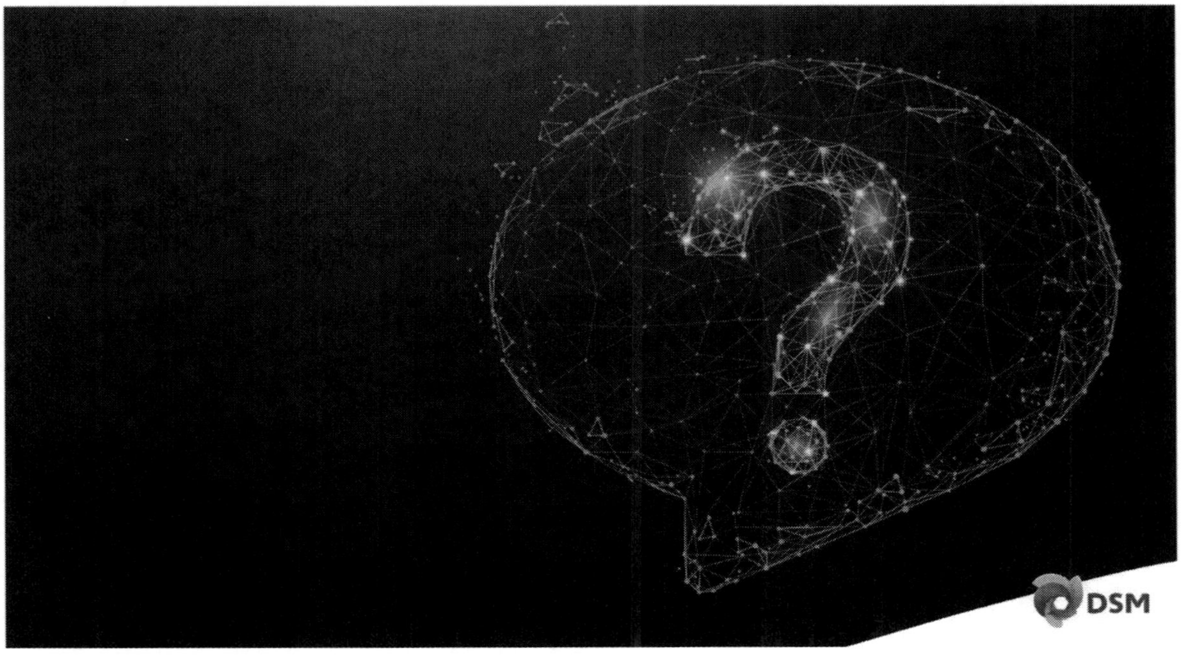

Purpose Led, Performance Driven
Powered by Digital
DSM

TOTAL DIGITAL FACTORY
14th of October 2020
2020 EUROPEAN CONFERENCE
ON PROCESS SAFETY & BIG DATA
CCPS EPSC
TOTAL

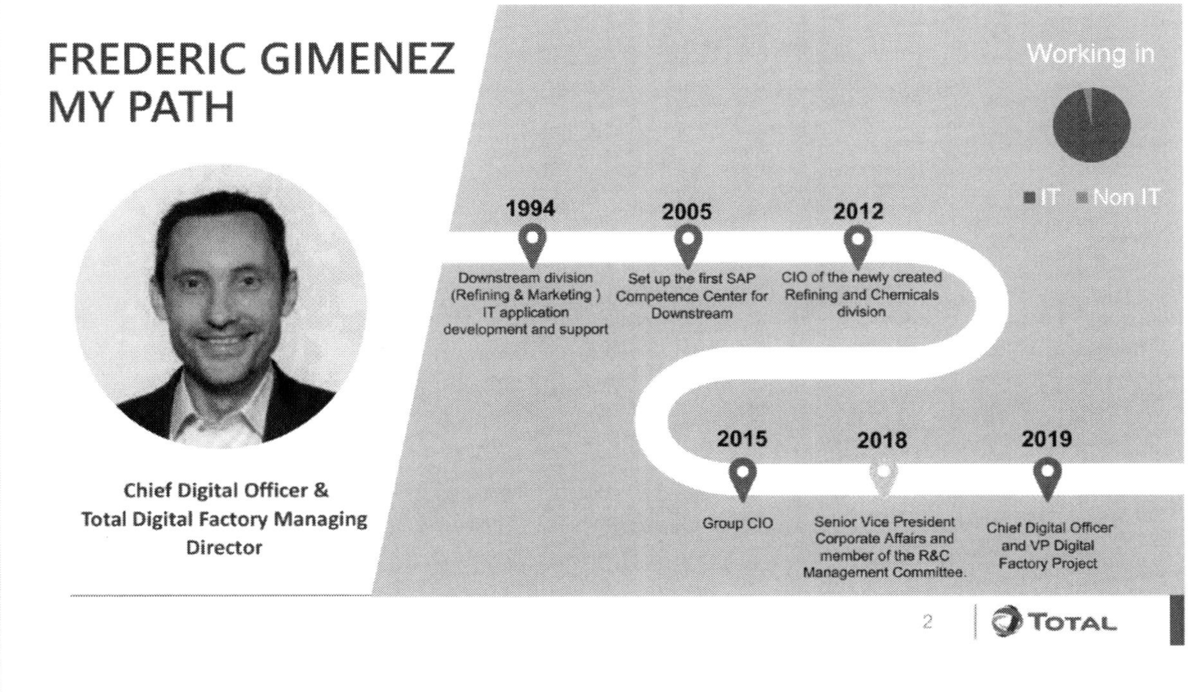
FREDERIC GIMENEZ
MY PATH
Chief Digital Officer &
Total Digital Factory Managing
Director
Working in
IT Non IT
1994
Downstream division (Refining & Marketing ) IT application development and support
2005
Set up the first SAP Competence Center for Downstream
2012
CIO of the newly created Refining and Chemicals division
2015
Group CIO
2018
Senior Vice President Corporate Affairs and member of the R&C Management Committee.
2019
Chief Digital Officer and VP Digital Factory Project
2
TOTAL

# TOTAL'S KEY FIGURES (2019)

**11.8 bn $** adjusted net income in 2019 - best-in-class profitability among majors

**>25 GW** production capacity of **renewable electricity** by 2025

**The world no. 2** liquefied natural gas operator (LNG)

**1.5-2 bn $** invested in **low-carbon electricity** each year

**3 mboe/day** of production in 2019, of which > **50% natural gas**

**6 major** integrated refining & petrochemical **complexes**

A global top **10 refiner** and **petrochemical** manufacturer

More than **8 millon** customers served in our **15,000 service stations** each day

**No. 4 lubricant retailer** in the world

**135 millon tons** of **crude** & **refined oil** shipped by sea

More than **4,000 researchers** in our **18 R & D** centers

**~1 bn $** spent on **R & D** in 2019 including **40% on low-carbon**

3 | TOTAL

## TRANSFORMING TOTAL INTO A BROAD ENERGY COMPANY

### OUR AMBITION: GETTING TO NET ZERO BY 2050

Natural Gas
- Grow LNG (#2 player) and develop biogas / clean $H_2$
- Promote natural gas for power and mobility

Electricity
- Accelerate investments in low carbon electricity primarily from renewables
- Integrate along the electricity value chain (production, storage, trading, supply)

Oil
- Focus investments on low cost oil and biofuels
- Adapt refining capacity and sales to demand in Europe

Carbon Sinks
- Invest in carbon sinks (NBS and CCUS)

**Total becoming Total EnergieS**

29 | TOTAL

# TOTAL'S AMBITION IN DIGITAL

A value-driven ambition with a target set at

**$ 1.5 Bn** per year in 2025

In value generated

| 1Bn$ **Upstream** | 0.5 Bn$ **Mid / Downstream** |
|---|---|
| • Production up 2%<br>• OPEX down 5%<br>• CAPEX down 5%<br>• Studies accelerated by 6 months | • Refining > $2 / T<br>• Petrochemical > $6 / T<br>• New mobility services for non-fuel: + $100Mn<br>• Support customer base growth ambition to 8 millions<br>• Optimized operations for GRP: + 200 M$ |

***+ HSE & CO2 performance***

TOTAL

# WHAT ARE WE TALKING ABOUT

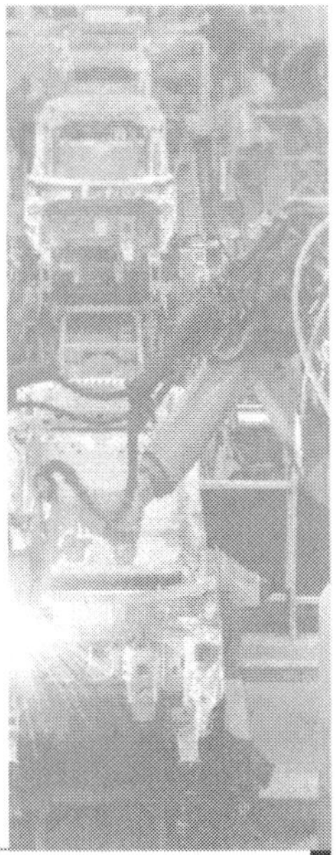

**DIGITAL**
**= Tech company**

- **Agility**
- **User-Centric**
- **Tech-focused**
- **Start-up mindset**

**+**

**FACTORY**
**= A production Unit**

- **Discipline**
- **Capacity to scale**
- **Delivery driven**
- **Robust organization**

TOTAL

# OUR MISSION

DELIVER

**BY CREATING TAILOR MADE DIGITAL SOLUTIONS APPLIED TO VARIOUS DOMAINS**

CUSTOMIZED SOLUTIONS BUILT WITH VALUE CREATION IN MIND

ACCELERATE

**BY BEING AGILE AND PRODUCING AT SCALE FOR OUR SITES ALL OVER THE WORLD**

BIGGER SCOPE, BETTER IMPACT, FASTER TRANSFORMATION

TRANSFORM

**BY BEING A REAL CATALYST OF NEW WAYS OF WORKING**

THINK, ACT, WORK, DECIDE, COLLABORATE DIFFERENTLY

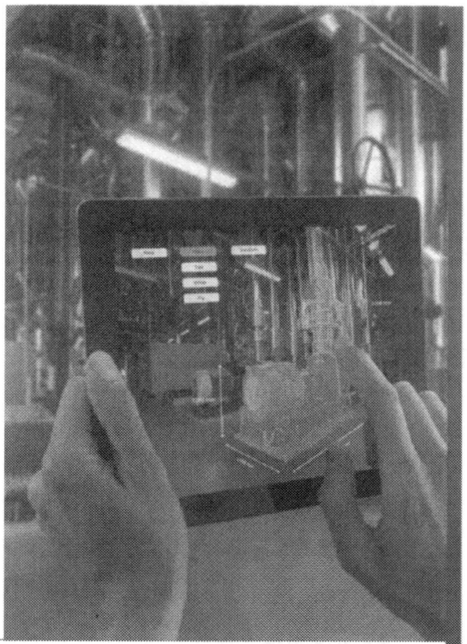



# OUR ORGANIZATION

Office of the CDO

Digital Transformation Office

Value & Portfolio | Talent & Academy | General Secretary

Data | Technical Authority | Digital Platforms

Delivery – Product squad

# OUR SQUADS ORGANIZATION

Delivery squads combine business expertise with a full range of digital skills.

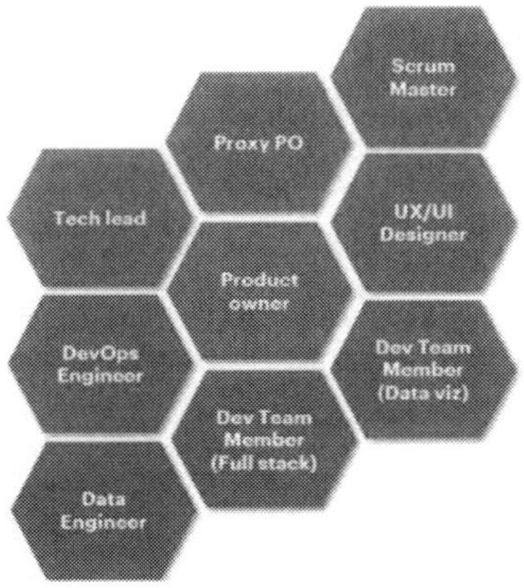

# OUR KEY FIGURES

***October 2020***

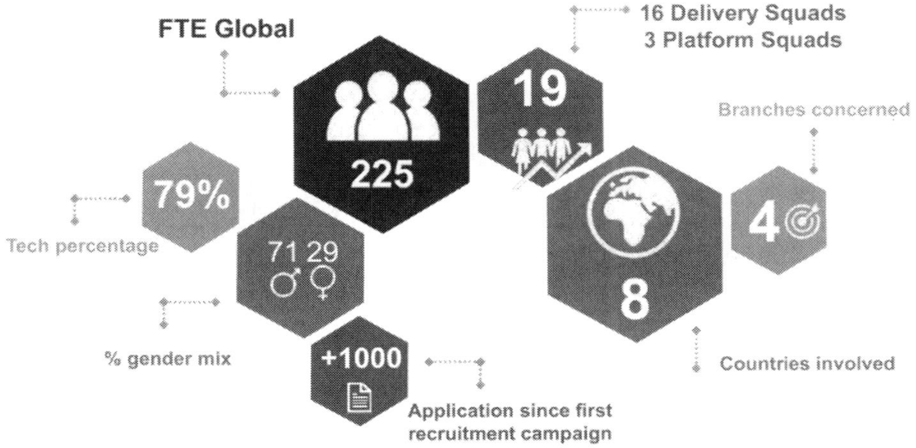

TOTAL

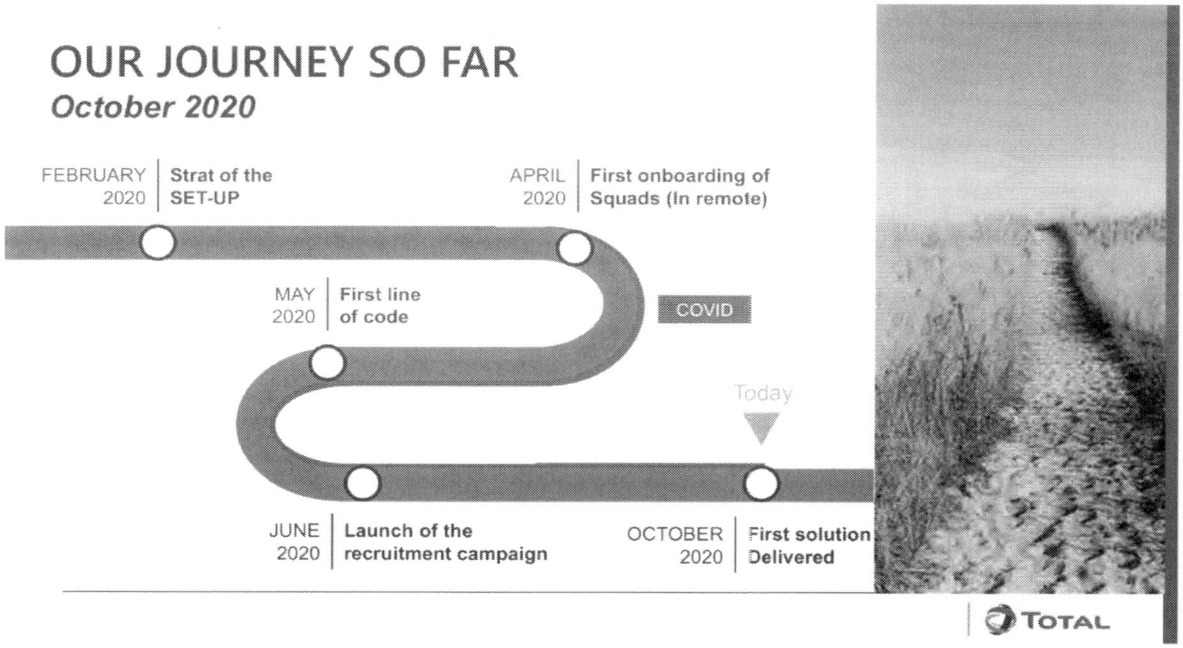
OUR JOURNEY SO FAR
October 2020
FEBRUARY 2020
Strat of the SET-UP
APRIL 2020
First onboarding of Squads (In remote)
MAY 2020
First line of code
COVID
Today
JUNE 2020
Launch of the recruitment campaign
OCTOBER 2020
First solution Delivered
TOTAL

PRODUCT LIFE CYCLE
12

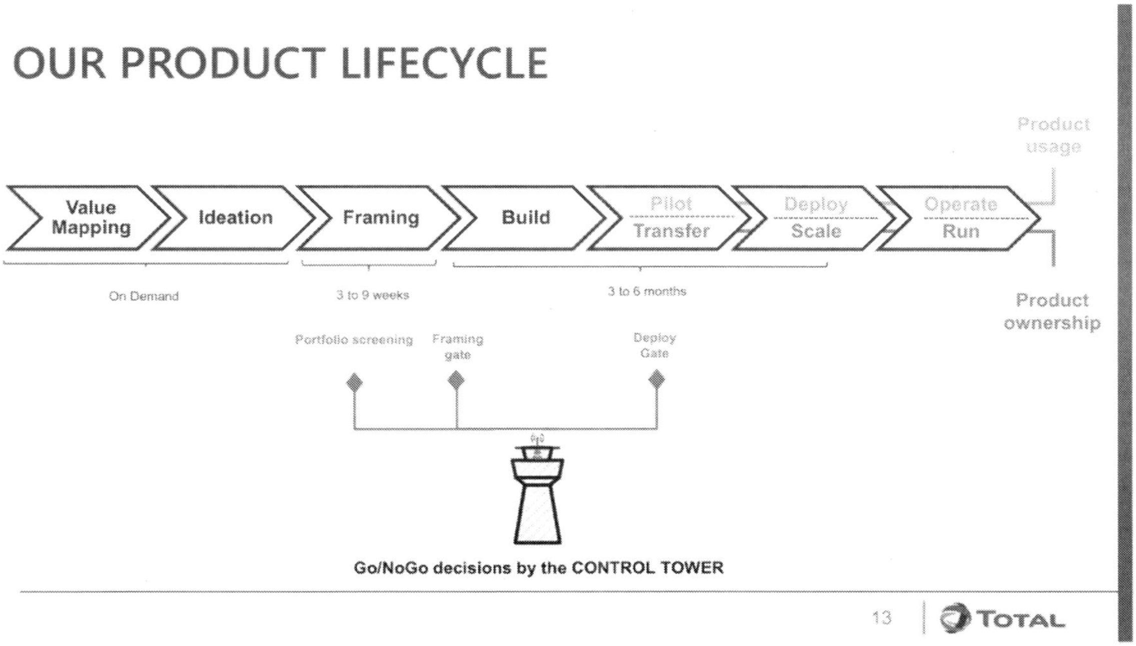
OUR PRODUCT LIFECYCLE
Product usage
Value Mapping
Ideation
Framing
Build
Pilot
Transfer
Deploy
Scale
Operate
Run
On Demand
3 to 9 weeks
3 to 6 months
Product ownership
Portfolio screening
Framing gate
Deploy Gate
Go/NoGo decisions by the CONTROL TOWER
13
TOTAL

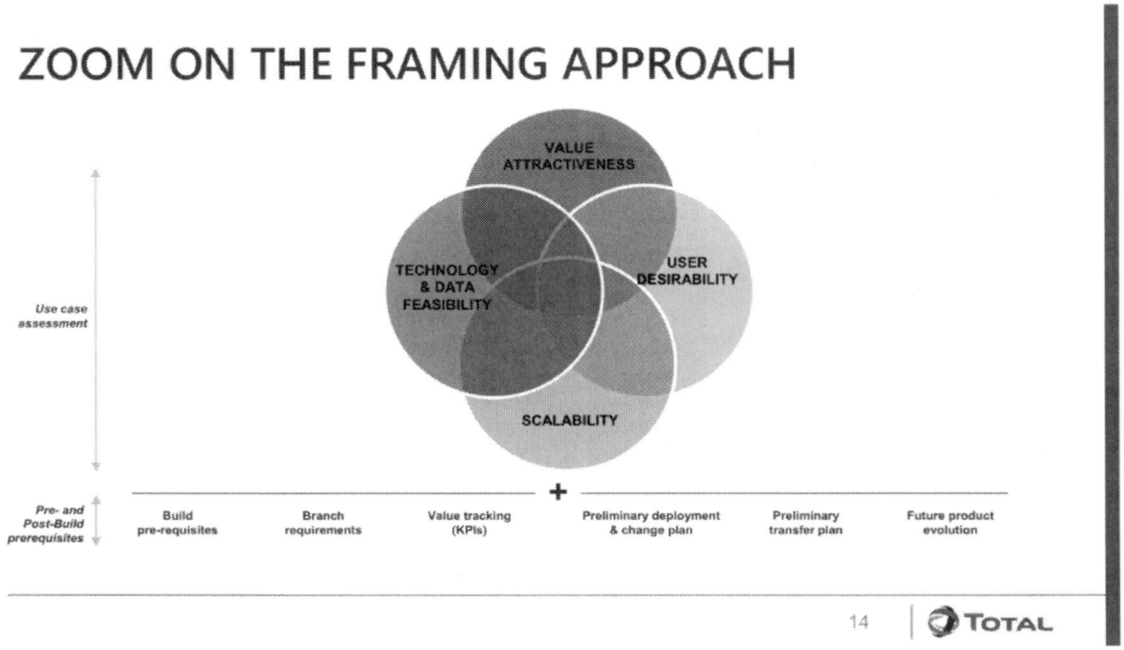
ZOOM ON THE FRAMING APPROACH
Use case assessment
VALUE ATTRACTIVENESS
TECHNOLOGY & DATA FEASIBILITY
USER DESIRABILITY
SCALABILITY
+
Pre- and Post-Build prerequisites
Build pre-requisites
Branch requirements
Value tracking (KPIs)
Preliminary deployment & change plan
Preliminary transfer plan
Future product evolution
14
TOTAL

# ZOOM ON THE BUILD METHODOLOGY

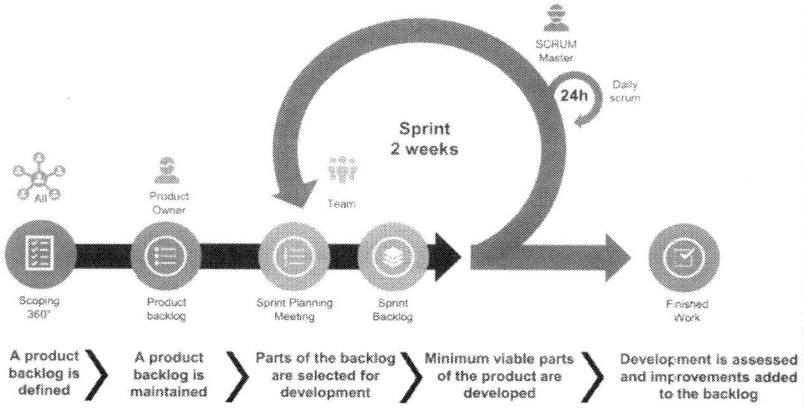

TOTAL

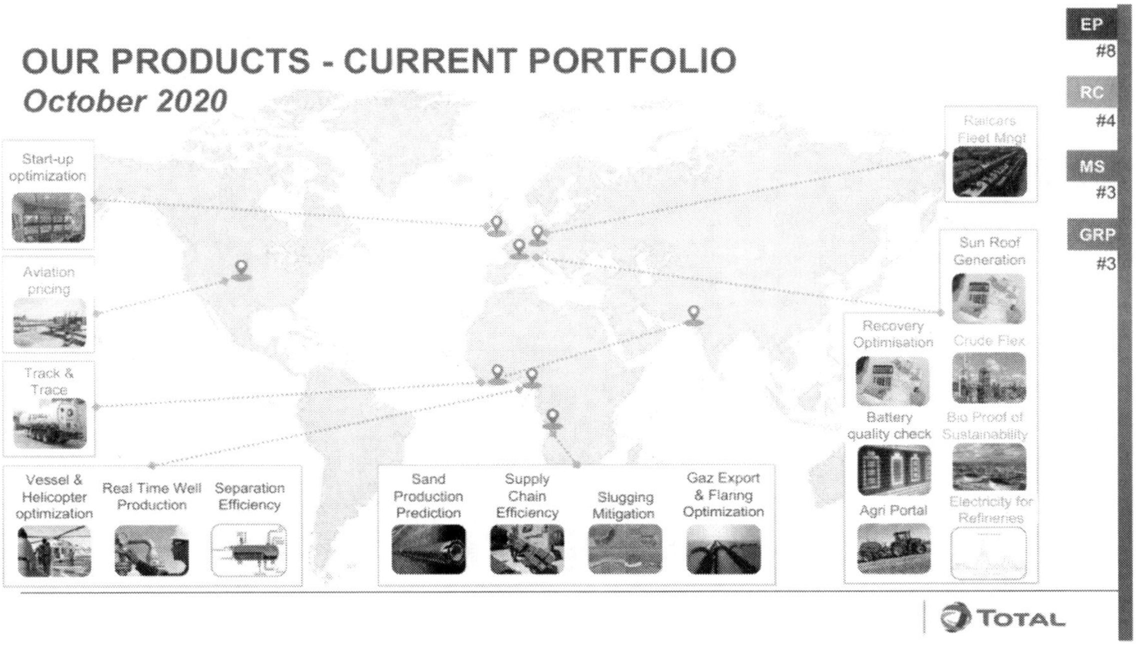
OUR PRODUCTS - CURRENT PORTFOLIO
October 2020
EP #8
RC #4
MS #3
GRP #3
Start-up optimization
Aviation pricing
Track & Trace
Vessel & Helicopter optimization
Real Time Well Production
Separation Efficiency
Sand Production Prediction
Supply Chain Efficiency
Slugging Mitigation
Gaz Export & Flaring Optimization
Railcars Fleet Mngt
Sun Roof Generation
Recovery Optimisation
Crude Flex
Battery quality check
Bio Proof of Sustainability
Agri Portal
Electricity for Refineries
TOTAL

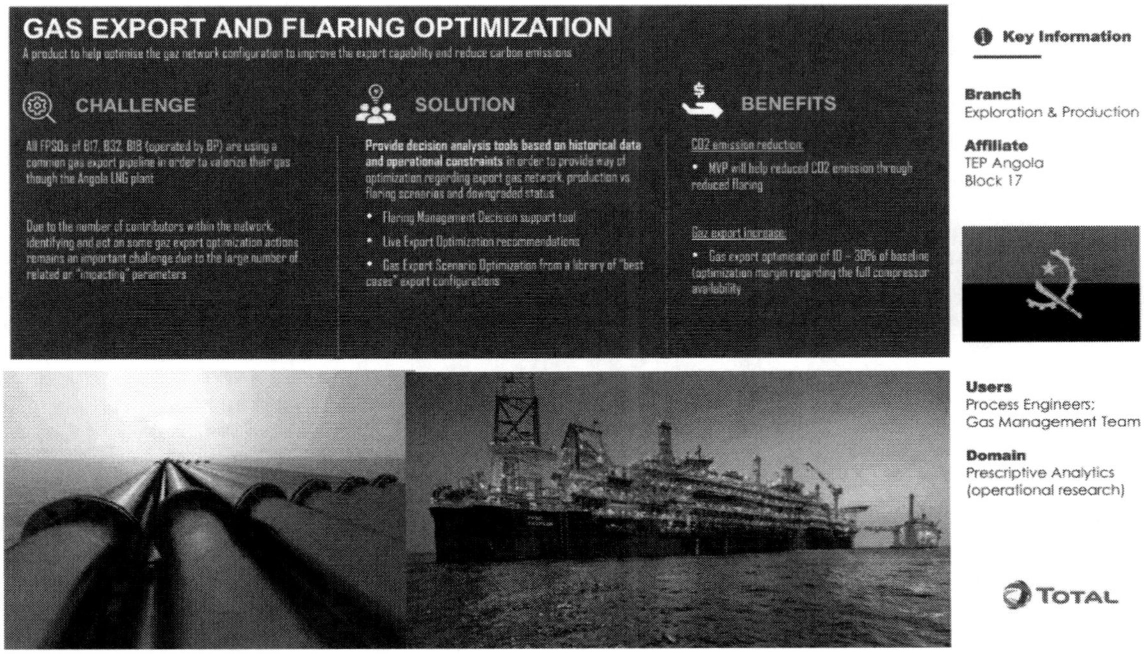
GAS EXPORT AND FLARING OPTIMIZATION
A product to help optimise the gaz network configuration to improve the export capability and reduce carbon emissions
CHALLENGE
All FPSOs of B17, B32, B18 (operated by BP) are using a common gas export pipeline in order to valorize their gas though the Angola LNG plant
Due to the number of contributors within the network, identifying and act on some gaz export optimization actions remains an important challenge due to the large number of related or "impacting" parameters
SOLUTION
Provide decision analysis tools based on historical data and operational constraints in order to provide way of optimization regarding export gas network, production vs flaring scenarios and downgraded status
• Flaring Management Decision support tool
• Live Export Optimization recommendations
• Gas Export Scenario Optimization from a library of "best cases" export configurations
BENEFITS
CO2 emission reduction:
• MVP will help reduced CO2 emission through reduced flaring
Gaz export increase:
• Gas export optimisation of 10 – 30% of baseline (optimization margin regarding the full compressor availability
Key Information
Branch
Exploration & Production
Affiliate
TEP Angola
Block 17
Users
Process Engineers;
Gas Management Team
Domain
Prescriptive Analytics (operational research)
TOTAL

## ELECTRICAL 4MATION

Reduce quality process lead time using data science techniques

### CHALLENGE

Quality process for electrical batteries takes nearly half of the overall manufacturing process.

It's currently based on an empirical approach given the high constraints on quality of products shipped to customers.

### SOLUTION

Data science engine that takes as input data collected along production lifecycle to predict batteries quality.

The solution requires a large shift in physical production lifecycle to handle data science specifics.

### BENEFITS

1. Reduce production lead time and therefore working capital
2. Increase customer satisfaction with shorter deliver-y timelines

**Key Information**

**Branch**
Gas Renewable & Power

**Affiliate**

saft a company of TOTAL

**Users**
Quality managers
Data managers

## BENEFITS OF TOTAL DIGITAL FACTORY

**Bring DIGITAL to assets**
and business organization

**Access to top digital TECHNO**
of Total Digital Factory

**UPSKILL the teams**
to new ways of working

Q&A
21

2020 EUROPEAN CONFERENCE ON PROCESS SAFETY & BIG DATA
October 13–14, 2020
CCPS EPSC

VIRTUAL

## PANEL 1

## *Customer Success Factors in Getting Started on Big Data & Analytics*

**Panel Moderators**

Dr. Abdul Aldeeb
Siemens Energy

Tony Downes
Honeywell

*Co-organized by the Center for Chemical Process Safety and the European Process Safety Centre*

2020 EUROPEAN CONFERENCE ON PROCESS SAFETY & BIG DATA
October 13–14, 2020
CCPS EPSC

VIRTUAL

## Customer Success Factors in Getting Started on Big Data & Analytics

**Panelists**

**Dr. Payam Jamshidi**, Siemens Energy, Global Lead for Mechanical Integrity

**Eugen Vlaicu**, Honeywell Connected Industry, Application Engineering Manager

**Joe Stough**, SIFZero Advisory Services, Founder

**Joep Coenen**, Versatec Energy, Sr. Consultant

*Co-organized by the Center for Chemical Process Safety and the European Process Safety Centre*

SIEMENS energy

# Leveraging Big Data for Better Business Decisions in Asset Integrity Management

**2020 European Conference on Process Safety & Big Data**

Panel 1: Customer Success Factors in Getting Started on Big Data & Analytics

Payam Jamshidi

13 October 2020



## Dr. Payam Jamshidi

SIEMENS energy

Payam Jamshidi is the Global Lead for Mechanical Integrity for Siemens Process & Safety Consulting. He is a chartered engineer with more than 12 years experience in Asset Integrity Management (AIM) with a core technical competency in Risk-Based Inspection (RBI) and Fitness-for-Service (FFS). He is a certified RBI trainer and a member of API 581, BS7910-Annex G and DS/1 for BS EN 16669.

Payam obtained his PhD in Materials Science and Engineering from University of London and completed his post-doc at the University of Manchester.

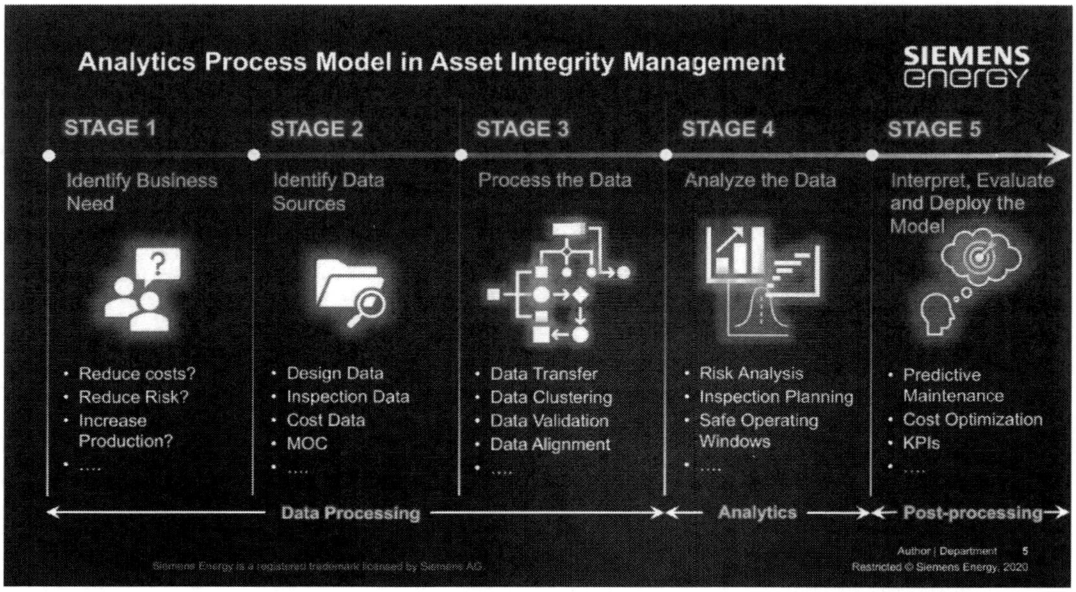
Analytics Process Model in Asset Integrity Management
SIEMENS energy
STAGE 1
STAGE 2
STAGE 3
STAGE 4
STAGE 5
Identify Business Need
Identify Data Sources
Process the Data
Analyze the Data
Interpret, Evaluate and Deploy the Model
• Reduce costs?
• Reduce Risk?
• Increase Production?
• ....
• Design Data
• Inspection Data
• Cost Data
• MOC
• ....
• Data Transfer
• Data Clustering
• Data Validation
• Data Alignment
• ....
• Risk Analysis
• Inspection Planning
• Safe Operating Windows
• ....
• Predictive Maintenance
• Cost Optimization
• KPIs
• ....
Data Processing
Analytics
Post-processing
Author | Department 5
Restricted © Siemens Energy, 2020
Siemens Energy is a registered trademark licensed by Siemens AG.

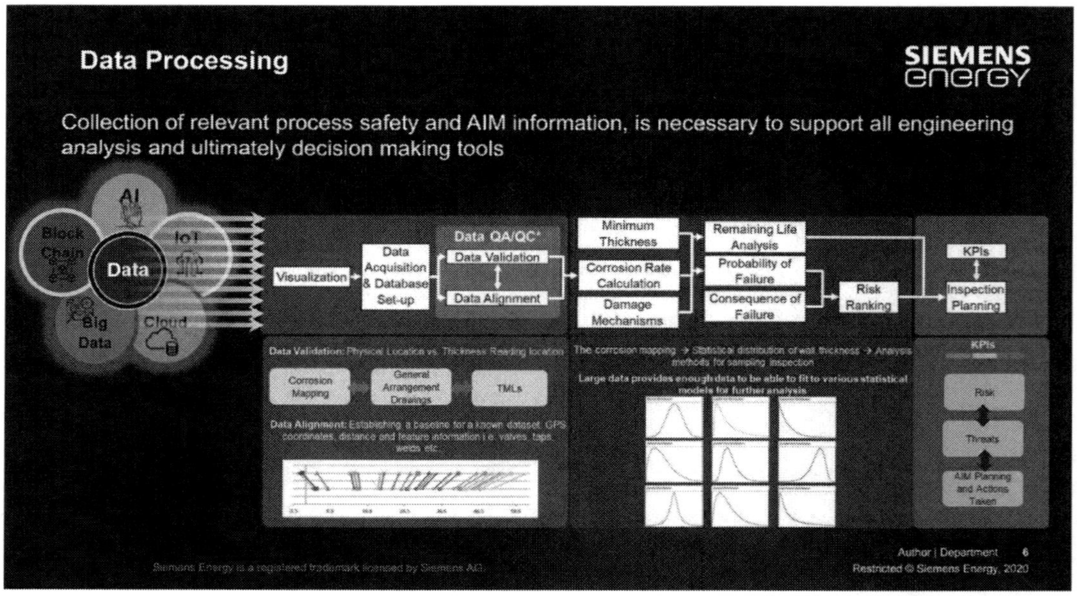
Data Processing
SIEMENS energy
Collection of relevant process safety and AIM information, is necessary to support all engineering analysis and ultimately decision making tools
AI
Block Chain
IoT
Data
Big Data
Cloud
Visualization
Data Acquisition & Database Set-up
Data QA/QC*
Data Validation
Data Alignment
Minimum Thickness
Corrosion Rate Calculation
Damage Mechanisms
Remaining Life Analysis
Probability of Failure
Consequence of Failure
Risk Ranking
KPIs
Inspection Planning
Data Validation: Physical Location vs. Thickness Reading location
Corrosion Mapping
General Arrangement Drawings
TMLs
Data Alignment: Establishing a baseline for a known dataset, GPS coordinates, distance and feature information i.e. valves, taps, welds etc.
The corrosion mapping → Statistical distribution of wall thickness → Analysis methods for sampling inspection
Large data provides enough data to be able to fit to various statistical models for further analysis
KPIs
Risk
Threats
AIM Planning and Actions Taken
Author | Department 6
Restricted © Siemens Energy, 2020
Siemens Energy is a registered trademark licensed by Siemens AG.

## Questions

SIEMENS energy

**Which stage of the data Analytics Process Model been your most challenging stage:**

a. STAGE 1: Identify Business Needs
b. STAGE 2: Identify Data Sources
c. STAGE 3: Process the Data
d. STAGE 4: Analyze the Data
e. STAGE 5: Interpret, Evaluate and Deploy the Model




## Overview of the Analytics Process Model

SIEMENS energy

**STAGE 1** Identify Business Need
- Reducing Uncertainty
- Meeting Health and Safety Requirements?
- Increase Asset's run-length between outages?
- Reduce Asset Outage Duration?
- Identify Mitigation Alternatives?
- Life Extension?
- Reduce Costs?

**STAGE 2** Identify Data Sources
- Design Data
- Operational Data
- Materials and Coatings
- Intrusive Inspection
- Thickness Surveys
- Conditional Monitoring
- Maintenance Records
- MOC records
- Repairs and Alterations
- Failure Records
- Climate/weather Records
- Project Cost Reports
- Plant Specific Failure Data
- Risk Analysis results

**STAGE 3** Process the Data
- Data Acquisition
- Data Virtualization Including:
  - Database Set-up
  - Cleaning the Data
  - Transforming the Data
  - Data Digitization and Organization (Clustering and Classification)
- QA/QC: Validation and Alignment
- Data Mapping and Modelling
- Visualization

**STAGE 4** Analyze the Data
- Defining Data Analytics Framework
- Sensitivity Analysis
- Degradation Mechanism Screening
- Failure Mode and Effect Analysis (FMEA)
- Inspection Frequency and Planning, i.e., Time, Condition, and Risk Based
- Risk Analysis
- Setting Safe Integrity Operating Windows

**STAGE 5** Interpret, Evaluate and Deploy the Model
- Portal Corrosion Monitoring and Inspection Planning Visualization
- Predictive Maintenance
- Monitoring KPIs
- Cost Risk Optimization
- Regression Analysis

← Data Processing → ← Analytics → ← Post-processing →

# EUGEN VLAICU

**Company:** **Honeywell Connected Enterprise**

**Current Role: Data & Analytics Engineering Manager**

**Current Location: Bucharest, Romania**

**Current Key Responsibilities:**

- Engineering team leader in EMEA region
- Consultancy
- Innovative solutions design

**Area of expertise:**

- Data acquisition, aggregation, processing
- Equipment Performance
- Energy Efficiency

**Experience:**

- 17 years in Engineering field
- 10 years in Data & Analytics

**Something about myself:**

- I like traveling, riding bike, fishing, bowling, going to rock concerts and festivals
- Travelled & lived in multiple countries
- Favorite place for going out: Manhattan – New York, US
- Favorite place to live: Malaga – Southern Spain

 **Honeywell**

# CHALLENGES OF MODERN INDUSTRY

## OPERATIONAL

Fragmented Views – islands of data and disconnected visibility
- ✓ Unify all plant data sources into a single data lake

Profit Loss – harder to improve after years of Lean or Six Sigma
- ✓ Implement equipment performance & energy monitoring tools

Capacity Constraints – unplanned downtime and bottlenecks
- ✓ Deploy PdM tools with digital twins and ML analytics

Safety Risks – constantly seeking latent EHS hazards
- ✓ Use ML and predict
threats

## HUMAN CAPITAL

Experience Loss – retirement, high SME demand
- ✓ Deploy an user advisory system to embed SME knowledge

Juggling Assets – struggle to manage all equipment
- ✓ Aggregate all common alarms into a single view application

Manual Calculations – utilizing multiple offline programs
- ✓ Use automatic / online template-based tools & equip. monitoring apps.

Run-to-Fail – where field data, experts or monitoring do not exist
- ✓ Implement inspection round tools to semi-automate the monitoring

 **Honeywell**

## QUESTION

1. Which of the below is the most significant challenge that you face currently
   A. Missing or islands of data
   B. Productivity improvement & unplanned downtime
   C. Latent EHS hazards
   D. Lack of subject matter expertise or struggle to manage all systems
   E. Missing monitoring or using offline tools (e.g. Excel)

 Honeywell

Joe Stough, Founder of SIFZero

Over 20 years founding and running enterprise software and analytics business in the Safety and Operational Excellence market. Recently founded SIFZero to deliver advisory services in same space.

## A predictive analytics approach to solve the SIF problem

SIFZero Advisory Services, LLC | Joe Stough - Oct 13, 2020 | www.linkedln.com/in/joe-stough

## Predictive Analytics Approach to Solve SIF Problem

**Define the Problem:**
- The Outcome variable Y (incident rate or SIF rate) is greater than ZERO.

**Known solution: if safeguards (the ideal X) are perfect, then Y = 0.**
- But difficult / impractical to holistically measure as predictor of Y.

**So, what measurable efforts (X) are in place to solve the problem (i.e. affect Y incident rates)?**
- Risk assessments, incident investigation, audits, observations, PHA, LOPA, FMEA, etc. to identify and rank safeguard weakness
- All above result in action items to repair weak safeguards
- Collectively the effectiveness of these efforts represent a company's safeguard improvement capability

**Hypothesis**
- Organizations (operating units within companies) that have better safeguard improvement capability will have incident rates closer to 0.

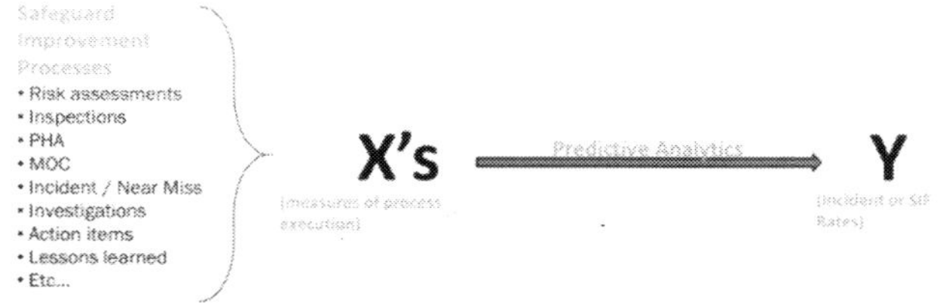


SIFZero Advisory Services, LLC | Joe Stough - Oct 13, 2020 | www.linkedin.com/in/joe-stough

## My Background: A Career-long Predictive Analytics Project

**Enterprise Data Platform**
- Incident investigations
- Audits, assessments, etc.
- PHA, MOC, etc.
- Preventive/corrective Action Items for ALL

**Multi-company Data Set**
- Over a million incidents (Y)
- Hundreds of total processes
- Thousands of operating assets
- Data set for analytics

**Predictive Analytics**
- Collaborative with large global co's
- Hundreds of metrics in data set
- Iterated to find predictive index
- Analytics per risk process maturity

**Evolving Measurements**
- Pragmatic to engage leadership
- Readiness / maturity-based

Findings per Predictive Analytics

**Evolving Areas of Measurement (X's) to Affect Outcome (Y's)**

1. **Culture**
   - Measure engagement in safeguard improvement processes
   - Measure discipline of responsiveness to identified gaps
2. **Materiality**
   - Measure reporting of low consequence (or near miss) with high potential
   - Measure resources skewed toward higher potential risks
3. **Quality**
   - Measure quality of investigations and findings (safeguard gap identification / ranking)
   - Measure safeguard efficacy

SIFZero Advisory Services, LLC | Joe Stough - Oct 13, 2020 | www.linkedin.com/in/joe-stough

# Question

- To what degree does your company engage in applying predictive analytics (machine learning) to identify / measure the factors that drive incident rate reduction?
    A. Little to none.
    B. We use predictive analytics for this purpose with data internal to our company only.
    C. We participate in multi-company predictive analytics benchmarking of leading and lagging indicators.
    D. Both B and C.

SIFZero Advisory Services, LLC    Joe Stough - Oct 13, 2020    www.linkedin.com/in/joe-stough

2020 EUROPEAN CONFERENCE ON PROCESS SAFETY & BIG DATA

October 13 – 14, 2020

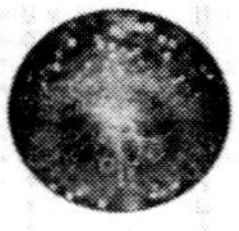

VIRTUAL

**Joep Coenen:**

- Sr. Consultant and MT-member @ Versatec Energy :

*Independent consultancy firm on safety, operational excellence and smart solutions*

- CTO and co-founder @ Hystream :

*Independent project developer on hydrogen solutions*

- Chairman "SIL platform" : User group of 50+ ind. members putting IEC 61511 functional safety into practice
- TüV certifed Functional Safety
- 20+ years experience in Oil & Gas, Chemical, Steel, Marine and Renewables
- Enthusiastic Waterpolo Player

*Joep Coenen*    *www.versatec.nl*

2020 EUROPEAN CONFERENCE ON PROCESS SAFETY & BIG DATA

October 13 – 14, 2020

CCPS EPSC

VIRTUAL

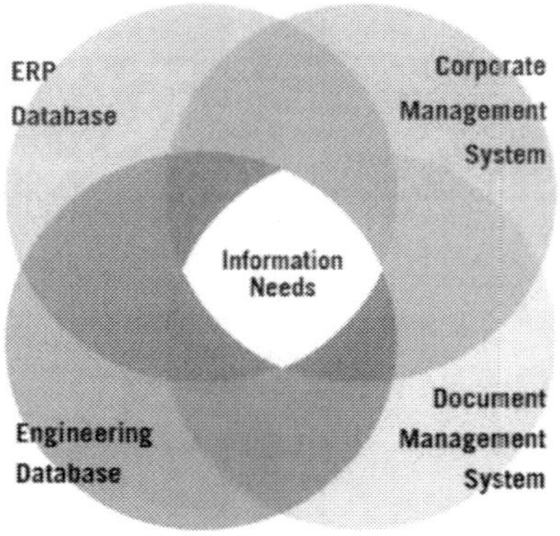


Joep Coenen

www.versatec.nl

2020 EUROPEAN CONFERENCE ON PROCESS SAFETY & BIG DATA

October 13 – 14, 2020

CCPS EPSC

VIRTUAL

## Data Management Maturity Model

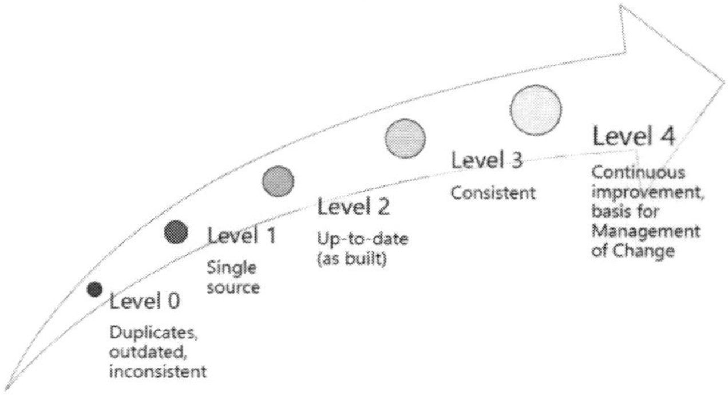


Joep Coenen

www.versatec.nl

2020 EUROPEAN CONFERENCE ON PROCESS SAFETY & BIG DATA
October 13–14, 2020
CCPS EPSC
VIRTUAL

## KEY QUESTION:

How does your company **proof compliance** in the area of Process Safety (PS) to the authorities given the fact that critical PS information is usually scattered across multiple data sources such as SAP, SharePoint, Smart plant PID, etc. ?

1. We **manually** search through the various data sources and collect PS data as and when requested **on the day** that the authorities are conducting their PS audit.
2. We **manually** search through the various data sources and collect PS data **in advance** to the PS audit.
3. We **manually** duplicate and update PS data from our data sources into a **single repository** to show during the PS audit.
4. We **automatically connect PS data** within different data sources with each other and ensure that the PS data is always consistent and complete every day.

*Joep Coenen* *www.versatec.nl*

2020 EUROPEAN CONFERENCE ON PROCESS SAFETY & BIG DATA
October 13–14, 2020
CCPS EPSC
VIRTUAL

## *Customer Success Factors in Getting Started on Big Data & Analytics*

## *Q&A*

*Co-organized by the Center for Chemical Process Safety and the European Process Safety Centre*

## Challenges to the Process Industry Operators for leveraging Big Data for Process Safety

### Panel Discussion

**Kiran Krishna**
*Systems Engineering Digital Lead, Shell Global Solutions (US) Inc*

**Olivier Peeters**
*Vice President Technology & Projects M&S - Total*

**Greg Nesmith, PE, CFSE**
*Senior Process Safety Manager for Industry 4.0 at Dow*

Moderator: Richard Pawulski, PE, P Eng
Vice President, Center for Technical Expertise - Air Liquide USA

1

# Richard Pawulski

Vice President, Center for Technical Expertise - Air Liquide USA

**Responsibilities**
Manage Technology Subject Matter Experts
Support North American Plant Operations

**Link to Process Safety and Big Data**
Supporting Air Liquide's Big Data initiative – Smart Innovative Operations (SIO)
Deployment of Air Liquide's Industrial Management System (IMS)
Managed corporate functions responsible for development and deployment of Process Safety Rules

**Affiliations**
Compressed Gas Association – Air Liquide company representative
Center for Chemical Process Safety - Air Liquide company representative

email: richard.pawulski@airliquide.com

2

# Kiran Krishna

Systems Engineering Digital Lead, Shell Global Solutions (US) Inc.

**Responsibilities**
- Product owner for Systems Engineering tools
- Leading industry efforts to move from documents to data in the Requirements Management

**Link to Process Safety and Big Data**
- Doctoral research at Mary Kay O'Conner Process Safety Center
- Past Technical Safety Engineering technical authority (TA2) for Upstream Americas
- Chair, CCPS sub-committee, "Bow Ties in Risk Management: A Concept Book for Process Safety"
- Center for Offshore Safety, developed and delivered first process safety indicators report for GoM
- Currently leading the selection, development and deployment of digital solutions for Systems Engineering

**Affiliations**
- Chartered Engineer - UK Engineering Council (CEng)
- Fellow of the Institution of Chemical Engineers (FIChemE)
- Fellow of the Institution of Mechanical Engineers (FIMechE)
- Professional Process Safety Engineer (via IChemE)

email: kiran.krishna@shell.com

3

# Digitization vs Digitalization

| Digitization | Digitalization |
|---|---|
| **Digitization** is the process of changing from analog to digital form.[1] | **Digitalization** is the use of digital technologies to change a business model and provide new revenue and value-producing opportunities; it is the process of moving to a digital business.[1] |
| • Developing plot plans and isometrics using software instead of drafting on paper. | • Enabling the latest approved plot plans and isometrics to be available on mobile devices at the construction site to reduce errors and re-work. |
| • Converting document based requirements into objects in a database. | • Transmission of requirements and verification to and from the supply chain to enable transparency requirement satisfaction and reduce quality issues. |

1 Gartner IT Glossary, http://www.gartner.com/it-glossary/

4

# Requirements Management

We ask for and get what we want – nothing more, nothing less – and we can prove it.

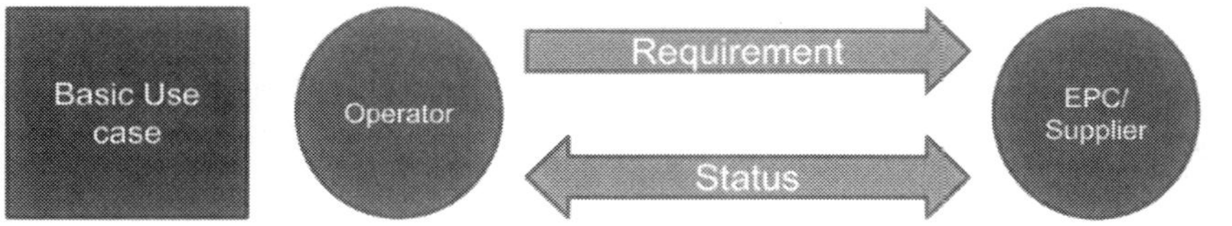


- Complexity increasing
- Documents per equipment growing
- Manhours per document increasing

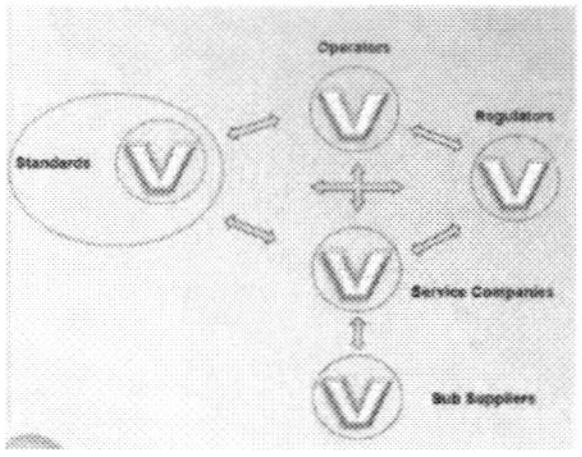


Complexity

Volume

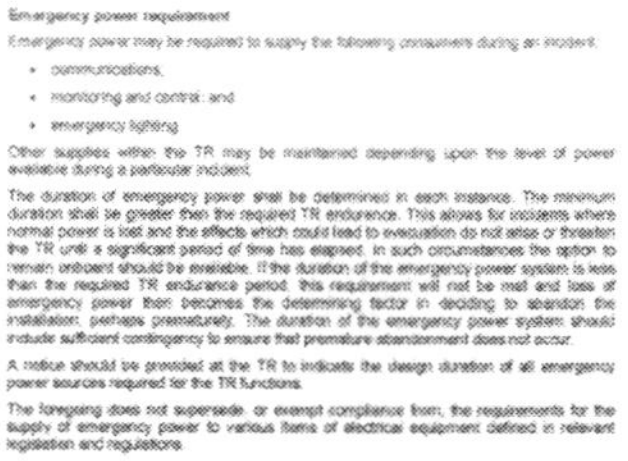

Emergency power requirement

Emergency power may be required to supply the following consumers during an incident:

- communications;
- monitoring and control; and
- emergency lighting.

Other supplies within the TR may be maintained depending upon the level of power available during a particular incident.

The duration of emergency power shall be determined in each instance. The minimum duration shall be greater than the required TR endurance. This allows for incidents where normal power is lost and the effects which could lead to evacuation do not arise or threaten the TR until a significant period of time has elapsed. In such circumstances the option to remain onboard should be available. If the duration of the emergency power system is less than the required TR endurance period, this requirement will not be met and loss of emergency power then becomes the determining factor in deciding to abandon the installation, perhaps prematurely. The duration of the emergency power system should include sufficient contingency to ensure that premature abandonment does not occur.

A notice should be provided at the TR to indicate the design duration of all emergency power sources required for the TR functions.

The foregoing does not supersede, or exempt compliance from, the requirements for the supply of emergency power to various items of electrical equipment defined in relevant legislation and regulations.

Ambiguity

5

# Requirements Management

We ask for and get what we want – nothing more, nothing less – and we can prove it.

## Data Centric Documents

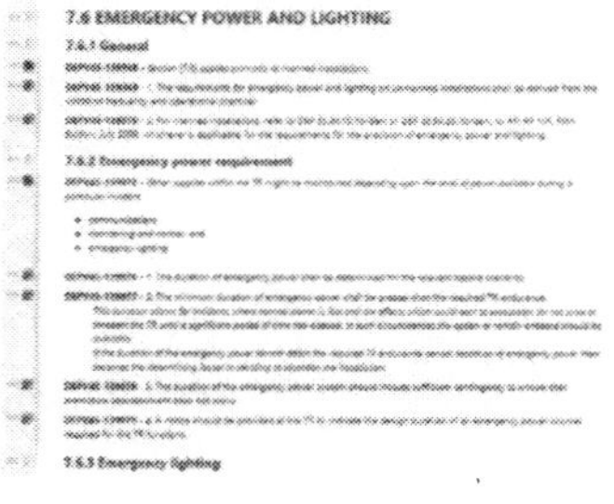


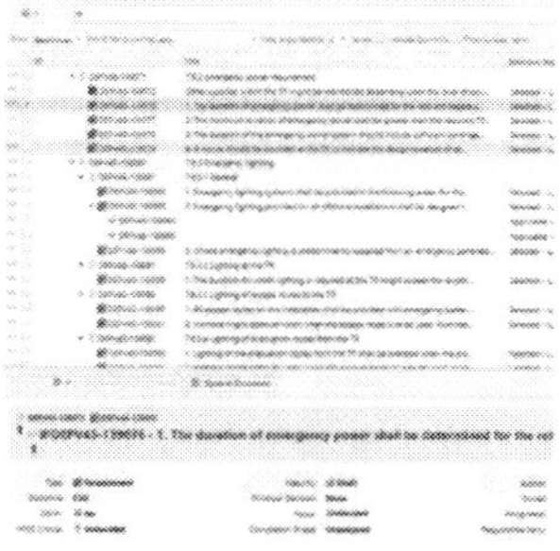

Toggles between a data view or a document view

## Traceability

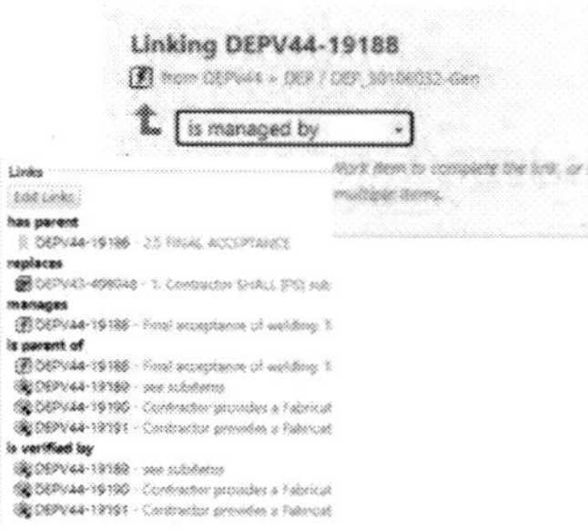


Data-centric mindset – Objects, metadata, and relationships

6

# Messages

## The Digitalization Triangle [2]

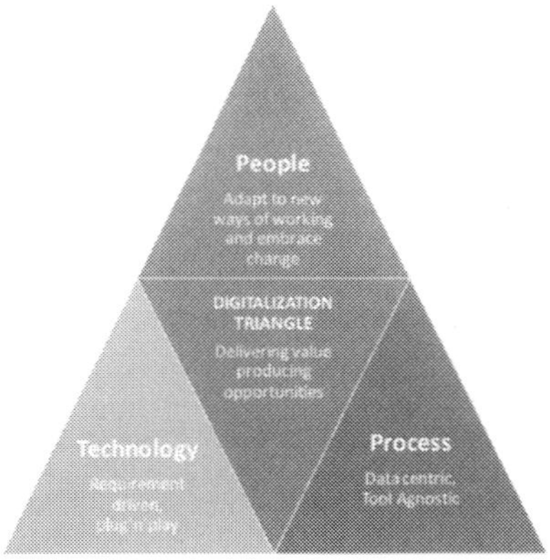


- Don't forget the users - not just about new toys
- User adoption will tell you a lot
- Standardize before you digitalize
  - Process – avoid silos
  - Data – data lakes or data islands
- Every digitalization initiative becomes a data clean-up exercise
- Measure value to help you fail fast
- Don't need a hammer to kill a fly – know your requirements

7

2. Anand, B., and Krishna, K., Overcoming Hurdles to Accelerate Data-centric Ways of Working in the Energy Industry, OTC-29446-MS, Offshore Technology Conference 2019, Houston, TX.

## Olivier PEETERS

Vice President Technology & Projects M&S - Total

**Responsibilities:**
In charge of industrial sites – from feasibility to construction

**Link to Process Safety and Big Data:**
Technologic Group Board Member
Cybersecurity executive committee

**CV Highlights**
- Master in Prevention (HSE)
- Electromechanical Engineer (Central School of Industrial Arts - Crafts)
- Lecturer "Ecole de Guerre", Paris.
- Expatriation - 15 Years: Africa - Europe

email: olivier.peeters@total.com

# Site Data Capture – Securing our Operations

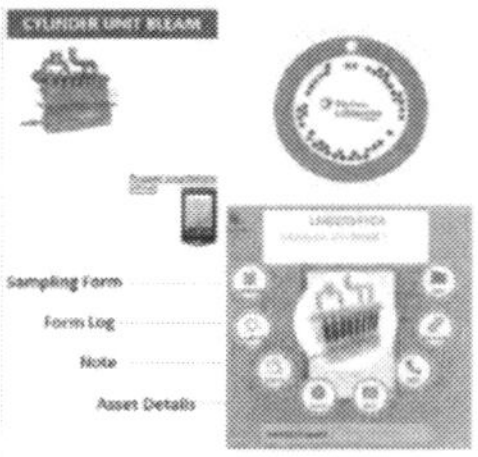


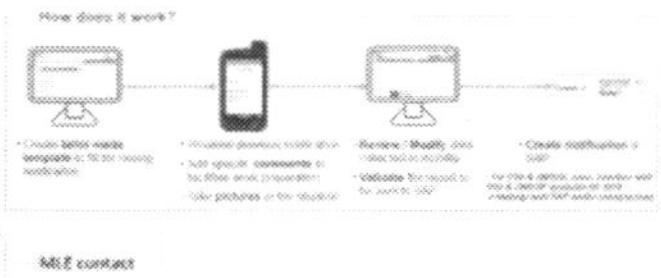


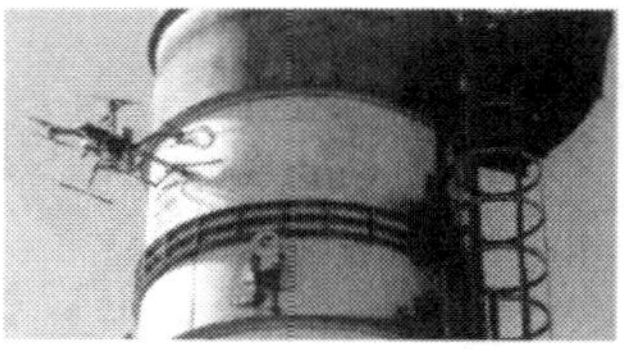

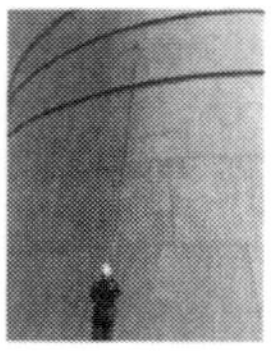

Total Industrial Mobility: .
to deliver mobile services for operational activities performed by any worker in a mobility situation on site.

**FlyNDT thickness measure**
(A-Scan, B-Scan, C-Scan)

**Leak detection using an acoustic camera**



TOTAL

# WORLDWIDE DRILLING REAL TIME SUPPORT CENTER (RTSC)

10

TOTAL

## Quantum is the **Digital Twin** of an installation (greenfield & existing asset)

Quantum is an **information management** program enabling instant visualisation of the installation with access to **all key up to date technical and operational information**



TOTAL

## INS Equipment - Transform **SAP Workorder** into **actionnable insights**

**Scope : 2005 ► 2019**
- **RN 184 970 work orders / 81 948 equipment**
- **UGO 587 408 work orders / 81 497 equipment**

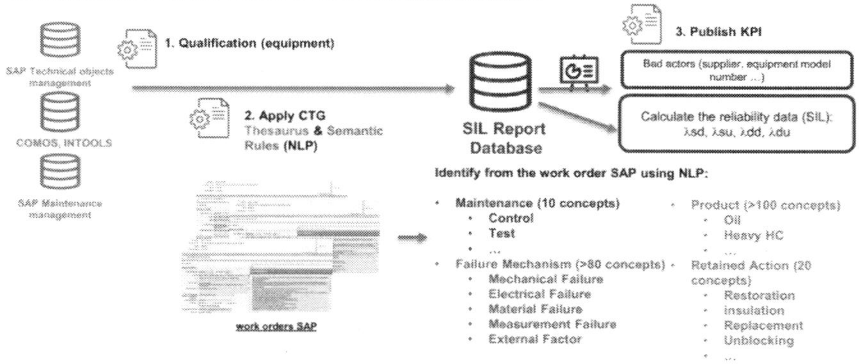






TOTAL

## Connected Helmet: Remote assistance for Improved Process Safety

**Examples** (successfully implemented)

- Startup inspection, plant commissioning
- Monitoring of the progress of execution works

on our construction sites

- Remote technical support despite the constraints

caused by the Covid-19 crisis

- Diagnosing operations & process issues
- Safety process review
- Remote Training– *under development*

## Connected Helmet: Remote assistance for Improved Process Safety

**Messages**

- A new step in our industrial digital transformation
- Digital Workforce and Expertise for Improved Safety Performance
- Remote support and technical assistance of our operations by experts
- Responsiveness, proximity and efficiency in monitoring Safety Process : It makes Total's operations safer and more efficient by executing better decisions faster
- Improve our operational excellence over the long run, and create the petrochemicals plant of tomorrow
- Support for the employees' skills development

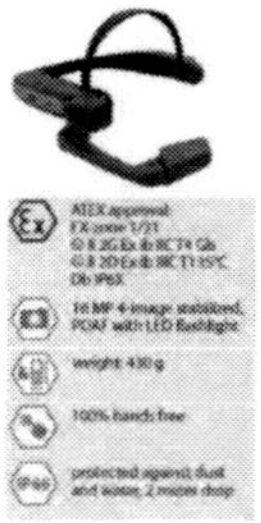

**Successes**

- Around 28 devices deployed by TOTAL MS & RC
- Continuity of technical support during the lockdown period

## SUMMARY

- The Fourth Industrial Revolution is a reality
- There is no doubt that there will be an impact on how technological risks will have to be managed
- Re-engineering of our current approach towards (process) safety is needed:
  - Automation of hazard identification studies and risk assessments
  - Use of advanced reality for training of operators
  - Use of digital in all safety management activities (work permits, standards, safety tours, safety meetings, learning, etc.)
  - Extensive use of robots for high risk tasks

**Management of Technological Risks = Good Engineering, Computing in a IoT & Digital world !!**



TOTAL

# Greg Nesmith, PE, CFSE

Senior Process Safety Manager for Industry 4.0 at Dow

**Responsibilities**
- Lead structured risk analyses
- Implement innovative solutions for Process Safety challenges
- Identify failure modes in new technologies

**Link to Process Safety and Big Data**
- Apply key instrument and automation standards (e.g., IEC 61511) in Dow
- Lead Data Analytics for SIS project

**Career Highlights**
- 30 years with Dow Chemical
- In Process Safety since 2007
- Certified Functional Safety Expert in 2017

email: gregnesmith@dow.com

PROTECTING People and Our Planet

DOW

## New Technologies = New Risks

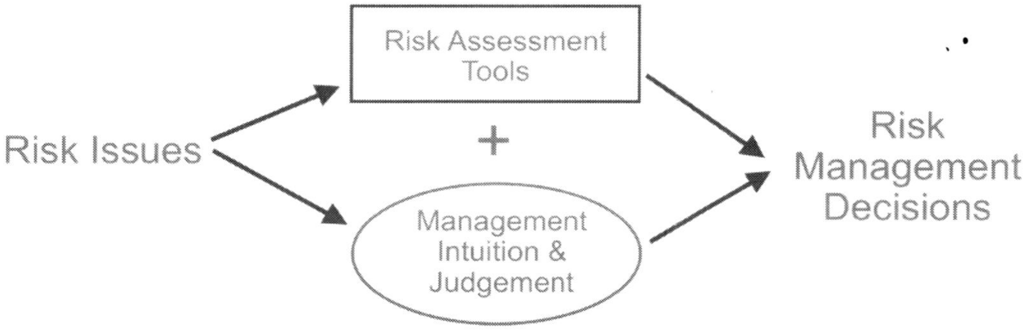


**Fundamental Challenges with New Technology:**

- Lack of historical data limits the tool set and effectiveness of tools
- Lack of experience hinders our intuition and judgment



## Hazards of New Technologies in a Chemical Manufacturing Facility

**Robots/Drones in Vessels:**

- HAC inside a vessel vs. outside
- "Clean" vessels (people vs. technology)
- Batteries do not have an "off" switch
- Static buildup on rotary blades
- Damage to internals
- Reactive chemical compatibility with device
- Pneumatic power (is air ok in this service?)

**Robots/Drones Outside of Vessels:**

- Dislodging hidden accumulated dust
- Passing through areas where flammables may be present

**Data Analytics:**

- Disruption of process automation
- Poor data
- Misapplication of data

## Example of a Successful Hazard Identification

**Hazards Identified:**

- Flammability (filling the sphere with water)
- HAC of nearby processes, namely other propylene sphere, temporary reclassification
- Cleaning process
- Decontamination of robot
- Drainage
- Gas detectors

**What else?**

- Water weighs more than propylene
- Check to see if civil design can handle extra weight

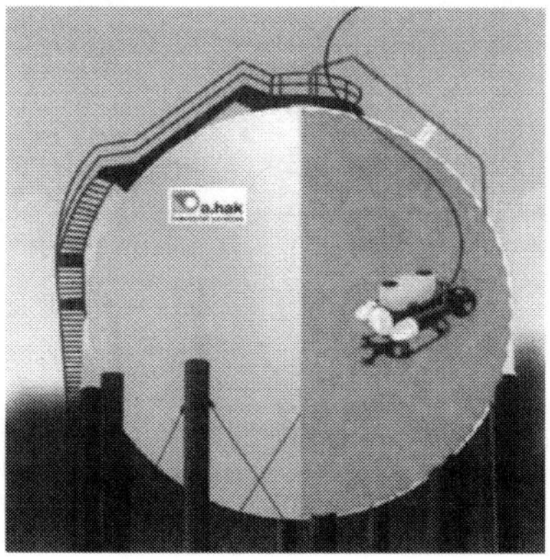

**Submarine Submerged Inside Process Vessel**
**Internal Inspection of Propylene Sphere (No CSE!)**

PROTECTING People and Our Planet

DOW



## Hazard Identification Tools

**Brainstorming Tools & Techniques**

- Failure Modes and Effects Analysis
- Hazard Operability Study (HAZOP)
- What-If Analysis
- Other techniques?

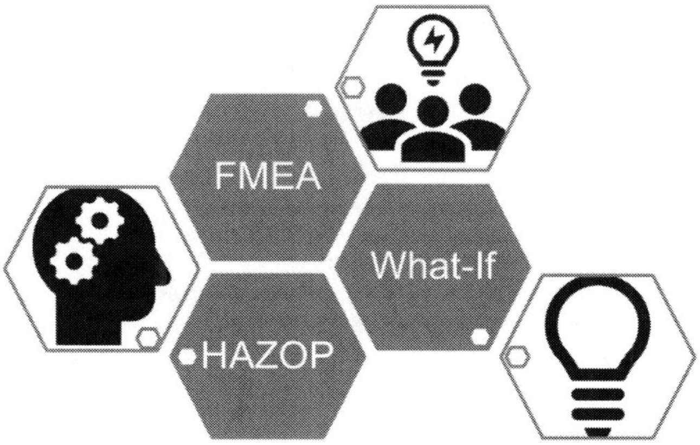


**Put together a winning team.**
**Emphasize quantity over quality.**
**No idea is too absurd to consider.**
**Use techniques that involve everyone.**
**All ideas become property of the team.**

PROTECTING People and Our Planet

DOW

sphera™

# Rethinking Process Safety & Risk Management in Operations

The Emergence of ORM Digital Twins

# Rethinking Process Safety & Risk Management in Operations

**The Emergence of ORM Digital Twins**

**SIMON JONES**
Sphera, Director
Solutions Consulting

**During today's presentation we will:**

- Review findings of 5th Annual PSM and ORM industry survey and gaps surfaced
- Explore the concept of digital twin and how it can be applied to connect process safety to frontline operations
- Look at data sources and risk models for constructing an ORM digital twin, drawing on an example implementation
- Look at the next steps on our ORM Digital Twin journey



sphera 2

## 2020 Process Safety & Operational Risk Mgmt Survey

5th Anniversary PSM/ORM Report

### GOOD INTENTIONS

Companies take safety seriously...

**88%** Say safety is part of corporate value structures

**78%** Continuously monitor safety performance

**60%** Are striving to reduce operational and Major Accident Hazard (MAH) risk exposure

... but gaps exist between intent and reality.

Say there are gaps between process safety intent and reality

Say risk changes during periodic process safety review periods



## 2020 Process Safety & Operational Risk Mgmt Survey

5th Anniversary PSM/ORM Report

Boldly going where no organization could go before

**3x**

growth in digital twin adoption

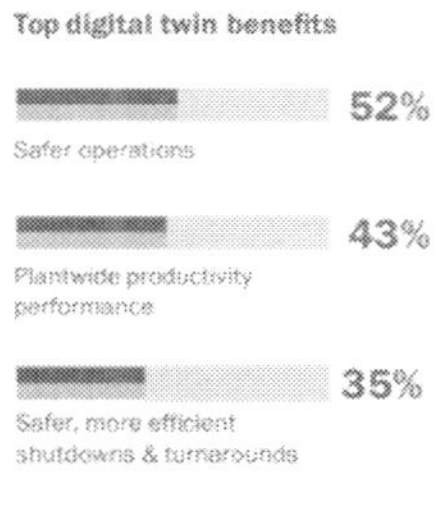

## Close the gaps: Operational Risk Management Digital Twin

**The "Physical" Asset** **The "ORM Digital Twin"**

*Improve Operational Decision-making*

- Understand
- Predict
- Act
- Monitor





## Close the gaps: Operational Risk Management Digital Twin

**The "Physical" Asset** **The "ORM Digital Twin"**

- Process Safety Critical Equipment status
- Overdue Safety Critical Maintenance
- Inspections
- Permitted Activity
- Operations Activity
- MoC
- Deviations
- Inhibits

*Improve Operational Decision-making*

- ► **Understand** how equipment status and activity all come together to impact cumulative risk
- ► **Predict** cumulative risk on the plant today & in the future
- ► Pro-**Act**ively manage productivity against risk
- ► **Monitor** the operational risk status of the asset

A true digital twin is a virtual counterpart of a real object, which enables other software, systems and operators to interact with it rather directly, bypassing the real object.

A digital twins must include:

- A model of the physical object
- Data generated by the object
- Unique one-to-one correspondence with the physical object
- The ability to monitor the object

## Constructing a digital twin: data sources

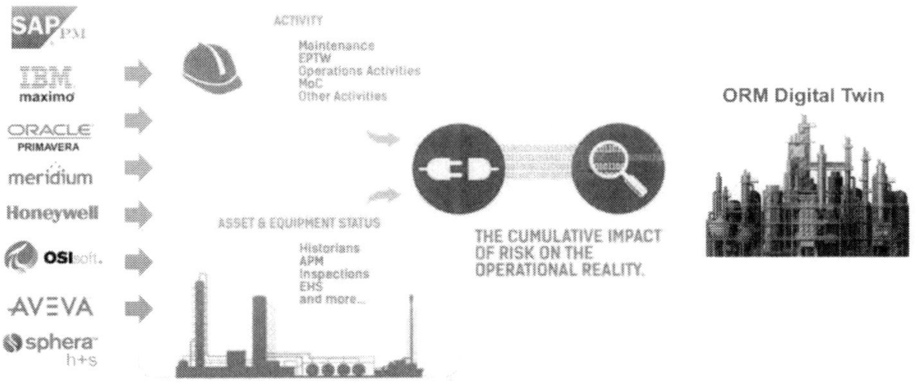


©2020 Sphera

sphera

## Constructing a digital twin: making risk visible

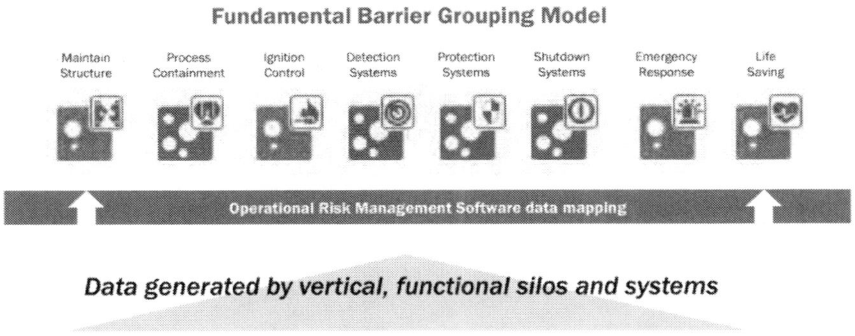


- Overdue Maintenance
- Overdue Inspections
- Planned Maintenance
- High Risk Jobs
- Failed Inspections
- Field Observations
- Management of Change
- Incident reports
- Critical Equipment Status
- Audit findings
- Shift competence gaps
- IOW Deviations

©2020 Sphera

Reference: IOGP - 544

sphera

## Close-the-loop between Operations, Maintenance & Engineering

CASE STUDY

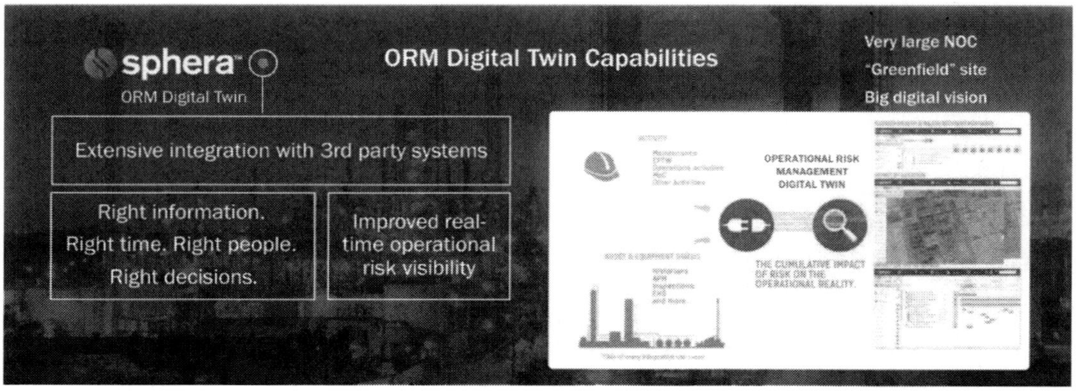


**Integrations**

© 2020 Sphera

sphera

## Another look at data sources

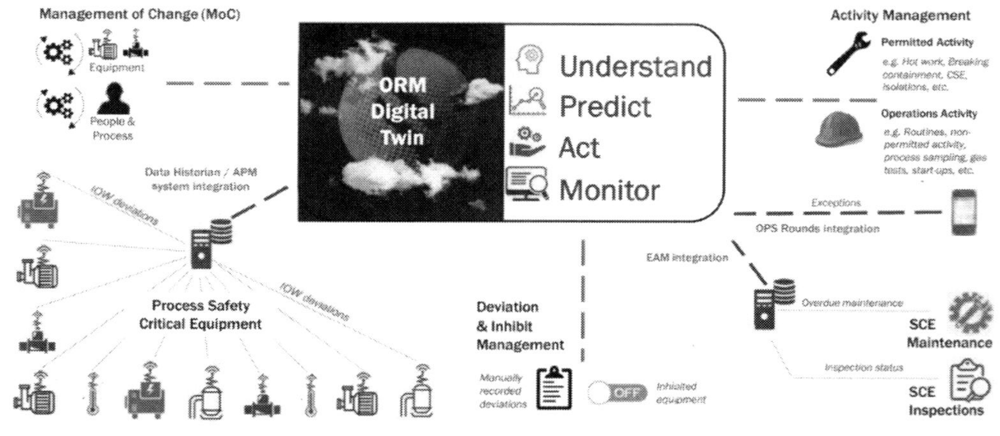


© 2020 Sphera

sphera

## Building on solid Process Safety foundation

Bowtie Diagrams

HAZOP & PHAs

LOPA

Safety Cases

**Real-time Capture of Risk related Data**

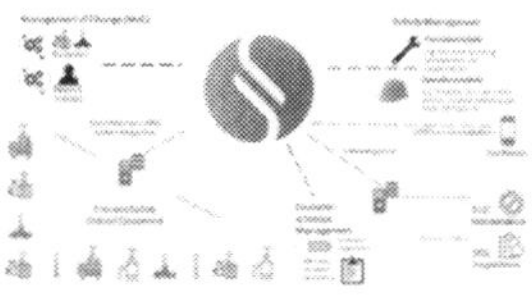

**Translation & Mapping to Barrier Model**

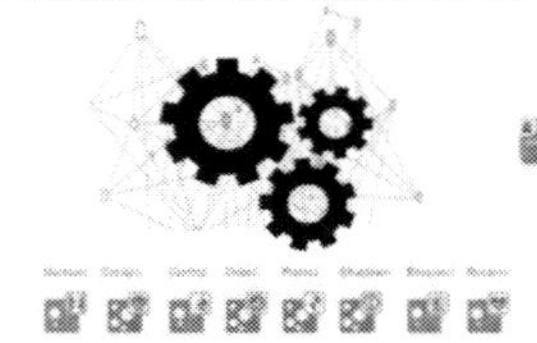

**Dynamic Real-Time Visualization**

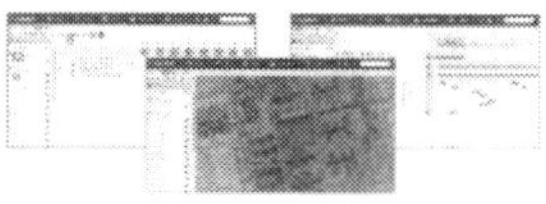

Frontline dashboards

*Drilldown*

Real-Time
Credible Risk Pathways

sphera 11

## Building a case – example: minimize unplanned downtime

ARC Advisory Group

- Total Installed Base of Systems Reaching the end of their Useful Life: $65 Billion
- Total Installed Base of Systems Older than 20yrs: $53 Billion
- Average Impact of Unscheduled Downtime: $20B or almost 5% of production in the process industries

2-5% of lost production in petro-chemicals

~$10,000/hr for a 2000 hp natural gas compression station

~10X the cost for unplanned events vs. planned maintenance in process industries

5-10% Increased inventories and labor costs, delayed delivery, reduced profitability in manufacturing



sphera 12

## How to Get Started

### Think big

Do it all from the start

Or

### Start small....

Integrated Ops. Mgmt – Control of Work with Activity Risk Management focus

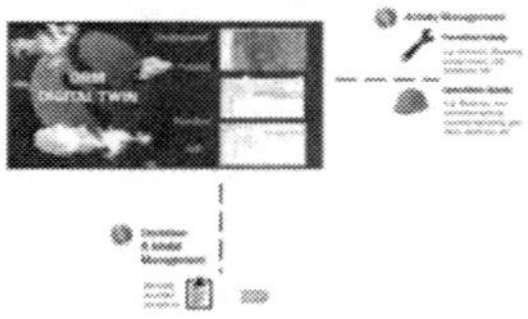

### ... scale fast

Add more capabilities

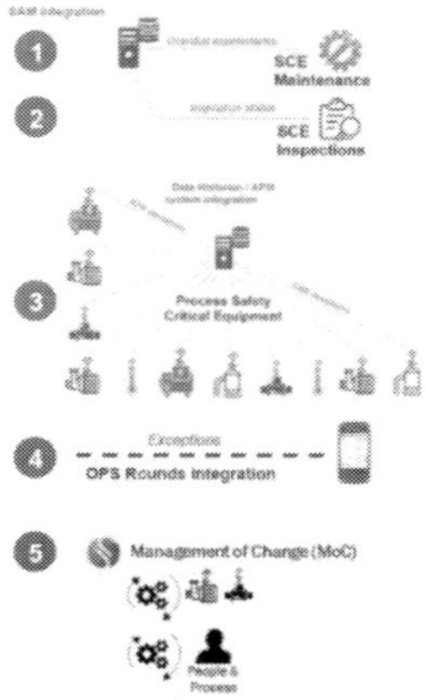






## Conclusion

**There are gaps between process safety intent and what happens at the frontline – the dynamic nature of frontline operations is a challenge**

Seize **your** opportunity to rethink process safety and risk management in operations – **key takeaways:**

- **Uncover the process safety-related data sources in your application environment today**
- **Build a business case and map out your journey – start out with destination in mind**
- **Use barrier health models to connect process safety to the frontline**

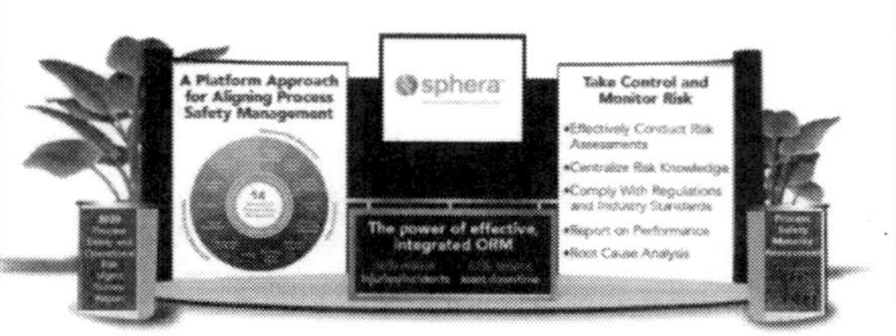


## Let's stay connected!

**SIMON JONES**
Sphera, Director
Solutions Consulting

Simon.Jones@sphera.com

KEY REFERENCES:

2020 Sphera Process Safety Survey report:
https://sphera.com/insights/2020-survey-on-the-state-of-process-safety-operational-risk-management/

IOGP 544 Standardization of Barrier Definitions:
https://www.iogp.org/bookstore/product/standardization-of-barrier-definitions/

Sphera Process Safety Maturity Model:
https://sphera.com/maturity-assessment-process-safety-management/



sphera™

## Three Screens | One Operational Reality

*Know what's happening, where it's happening, when it's happening, and what's driving the risk*

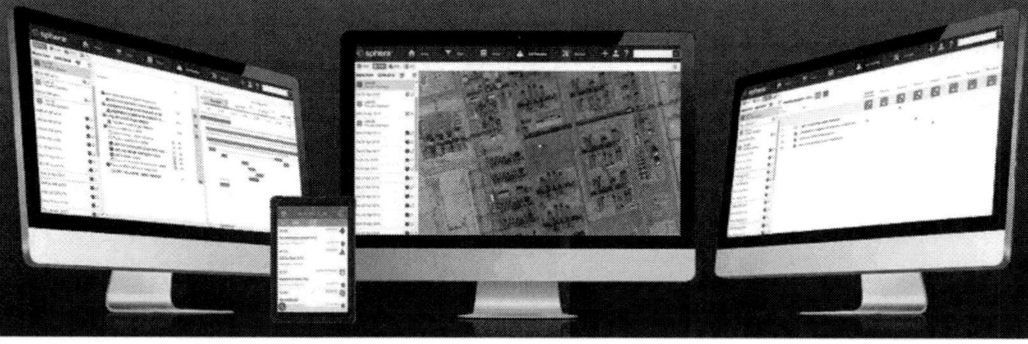

# Open source data mining and machine learning for process alarms of SCADA systems

Gyula Dörgő[1,2], Tamás Ruppert[2,3], Ahmet Palazoglu[1], János Abonyi[2]

[1]Department of Chemical Engineering, University of California, Davis, USA

[2]MTA-PE Lendület Complex Systems Monitoring Research Group, Department of Process Engineering, University of Pannonia

[3]Rabbit Miner Ltd., Veszprem, Hungary

Rabbit Miner | Complex System Monitoring Research Group | UC DAVIS COLLEGE of ENGINEERING

## Do we need operators?

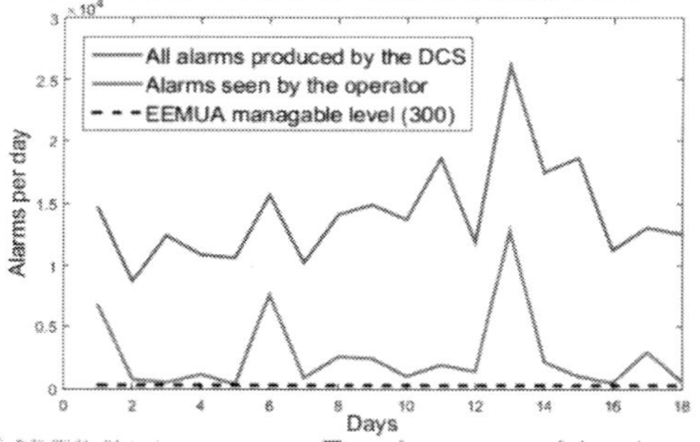


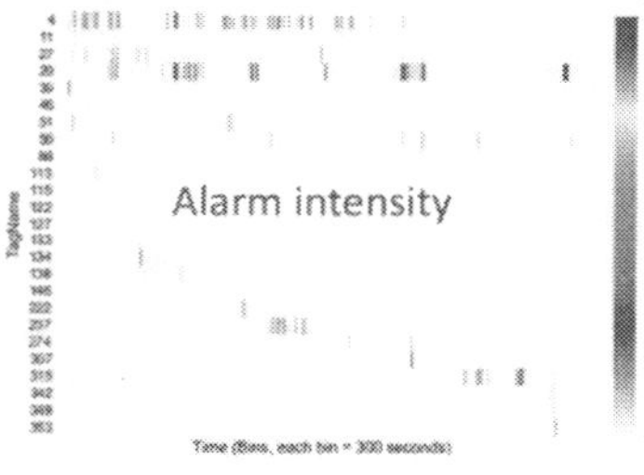


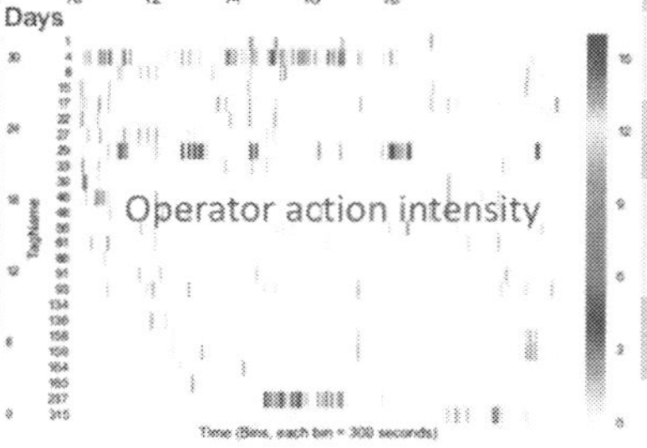


| Alarm performance metrics based on 30< days of data | | |
|---|---|---|
| Metric | Target Value | |
| | Acceptable | Max. |
| Annunciated alarms per hour per operator console | ~6 | ~12 |
| Annunciated alarms per 10 minutes per operator console | ~1 | ~2 |
| Percentage of 10-minute periods containing more than 10 alarms | ~<1 % | |
| Maximum number of alarms in a 10-minute period | ≤10 | |
| Percentage of time the alarm system is in a flood condition | ~<1 % | |
| Percentage of the top 10 most frequent alarms to the overall alarm load | ~<1 % to 5 % maximum | |
| Quantity of chattering and fleeting alarms | Zero | |
| Stale (standing) alarms | Less than 5 present on any day | |
| Annunciated priority distribution 3 priorities: (Low-Medium-High) | 80 – 15 – 5 % | |

1/17

# "Self-driving" operation

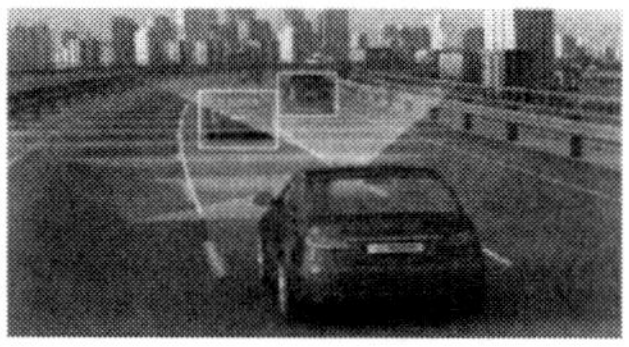

| Tasks | | Results |
|---|---|---|
| Exploring the alarm messages | → | Alarm load is decreased |
| Exploring the alarm-control action connections | → | Recommendation system |
| Validation of the control actions | → | **Fully automatized (and optimal) control** |



# Concept

We have expertise in building data-driven models to:

- detect the correlated alarm tags
- develop predictive models to make an event prediction
- make assumptions on the past and carry out a root cause analysis.

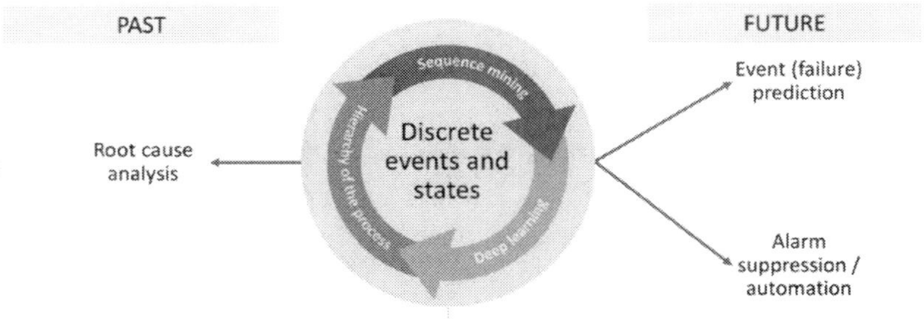


**Discrete events**
- State of the process
- Frequent sequence mining for prediction
- Alarm – operator interactions

**Deep learning**
- Root cause analysis
- Event prediction
- Visualisation for interpretability

**Hierarchy of the process**
- Structural process models
- Helps the freq. seq. mining
- Monitoring the spillover of malfunctions

## Sequence mining

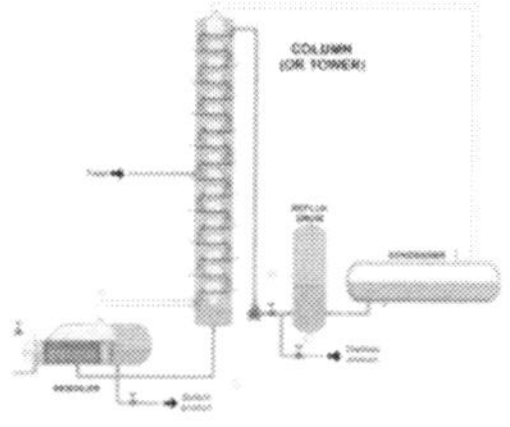

### Alarm messages

*Fouling*

↓

$F^{L}_{reflux} \Rightarrow p^{H}_{condenser} \Rightarrow T^{H}_{column\ top} \Rightarrow T^{H}_{condenser} \Rightarrow A^{L}_{product}$

### Operator actions

$Flow^{H}_{pump} \Rightarrow Q^{H}_{cooling} \Rightarrow Flow^{L}_{inflow} \Rightarrow Bypass$

Causal relations?

### Techniques

- Frequent sequence mining
- Deep learning

4/17

## The method and the framework

Open source machine learning tools are applicable in the process industry

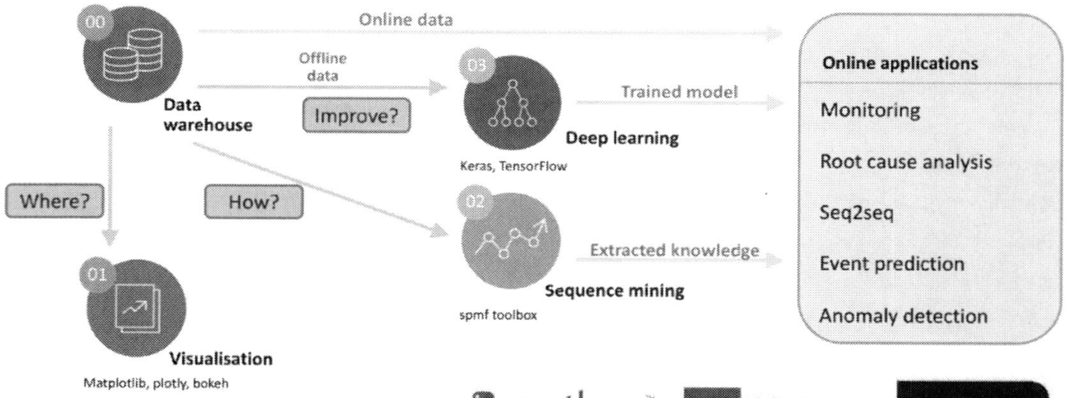


python Keras ProM

5/17

## The development process

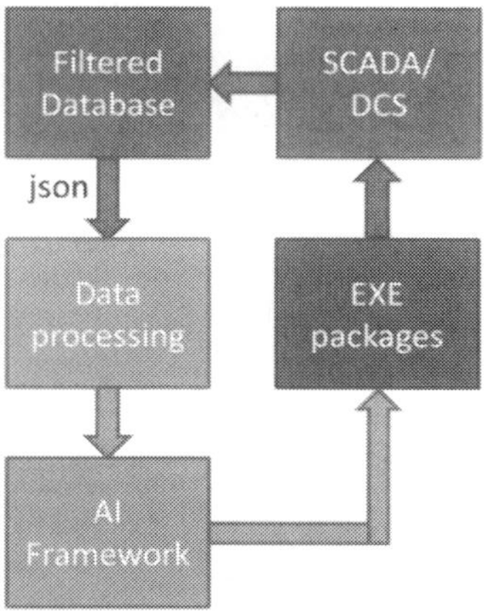


1 Business problem

2 Proof of Concept

3 Define requirements

4 Use-cases

5 GUI and integration

6 Test

6/17

## Alarm load

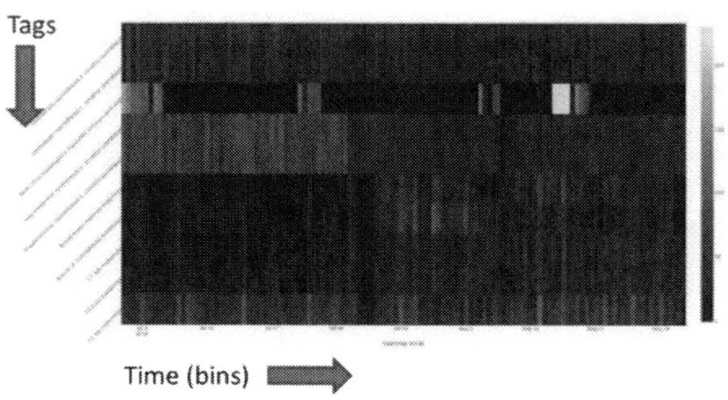


**Alarm intensity heatmap**

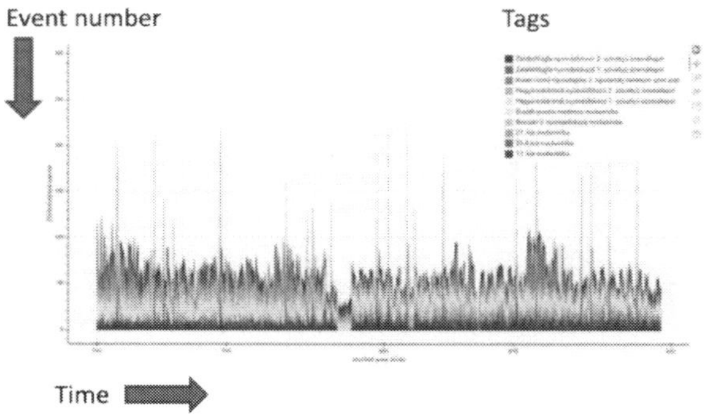


**Stacked area chart**

7/17

## Local workload of operators – Dotted chart

Tags/ operators/ units

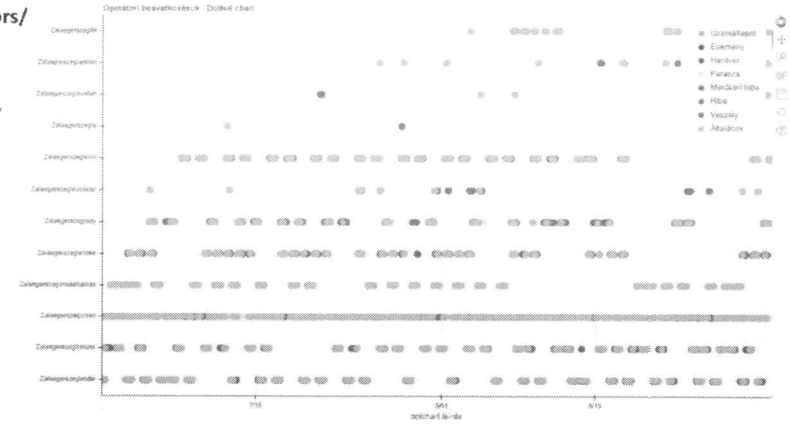

Time

**Colours:**

- Operating condition
- Event
- Hardware
- Command
- Control error
- Error
- Danger
- General

8/17

## Visualising the sequences – water system

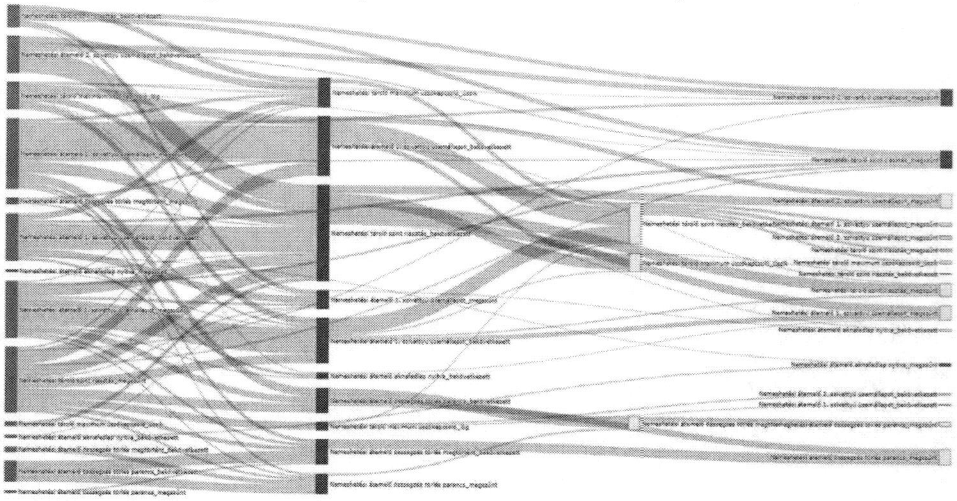

Sankey diagram - Proportional to support values (occurrences)

9/17

# webSCADA before/after

10/17

# Conclusions – act now!

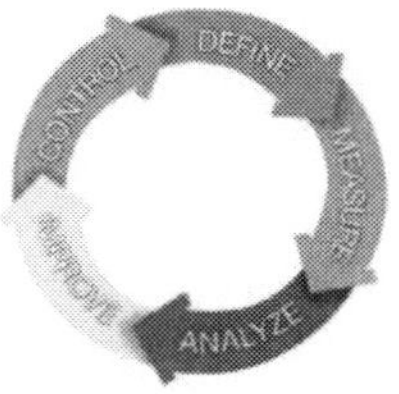


- The number of sensors is increasing!
- The number of operators is decreasing!
- The monitored data is unused – acquired, stored, managed...
- Costs?

**Tasks**

- Where is the problem?
- Sequence mining
- Event Prediction

Problem

**Realization**

- Visualization
- Advanced techniques
- Open source
- **Incorporation of process-relevant knowlege**

Solution

**Future**

- KPIs of operation actions
- Rare events?
- Recommendation system

Improvement

11/17

## Rabbit Miner – The process systems engineers

- Optimization
- Process systems engineering
- Process/Data mining
- Data science applications
- Industry 4.0 solutions

Contact:
Web: www.rabbitminerlab.com
E-mail: miner@rabbitminerlab.com

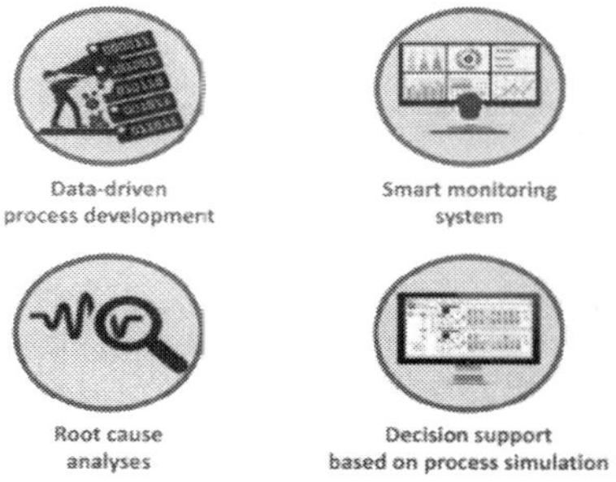


Rabbit Miner

12/17

## Thank You for your attention!

**The work has been supported by EFOP-3.6.1-16-2016-00015 Smart Specialization Startegy (S3) - Comprehensive Institutional Development Program at the University of Pannonia to Promote Sensible Individual Education and Career Choices**

**A kutatást támogatta a Nemzeti Kutatási, Fejlesztési és Innovációs Alap az OTKA–116674 pályázaton keresztül (Természettudományos kutatás és technológiafejlesztés folyamatbányászati és deep learning algoritmusokkal).**

**AZ EMBERI ERŐFORRÁSOK MINISZTÉRIUMA ÚJ NEMZETI KIVÁLÓSÁG PROGRAMJÁNAK TÁMOGATÁSÁVAL KÉSZÜLT**

13/17

Conf. Process Safety & Big Data,
Frankfurt, 10-2020

ProSafeX

atlan-tec
OPTIMIZE YOUR OPTIONS.

# HAZOP+
## Combining HAZOP and Operational Optimisation by Machine Learning

Hans V. Schwarz
Thomas Froese
10/2020

---

ProSafeX

## ProsafeX - Process Safety Excellence

ProSafeX

**Dr Hans V. SCHWARZ**
Process Safety Consultant

*ProsafeX*
Kronberger Str. 59
61462 Königstein,
Germany

+49 1520 921 5207
Hans.Schwarz@prosafex.com
Hans.V.Schwarz@outlook.de
www.prosafex.com

**Former VP Process Safety at BASF**

TÜV SÜD Chemie Service

ProSafeX
Process Safety Excellence

- *Management System*
- *Safety Hazard & Risk Assessment*
- *Incident assessment & Avoidance strategies*
- *Facility siting*
- *Digitalization of Safety and synergies with optimization of plant economics*

**Boardmember of:** EPSC THE PROCESS SAFETY NETWORK

Dr. Hans V. Schwarz, ProsafeX, 2020

## atlan-tec - Automated Machine Learning based Optimisation

atlan-tec
OPTIMIZE YOUR OPTIONS.

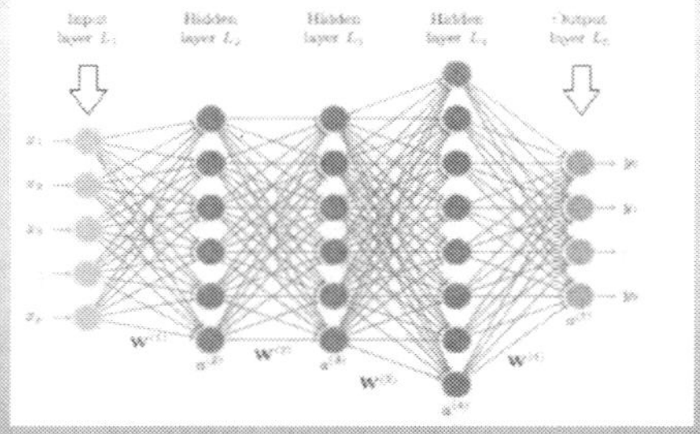

**Chairman of Standardization Committee "Big Data" FA7.24 //**
**Co-Chairman of Department 7:**

VDI VDE VDI/VDE-Gesellschaft Mess- und Automatisierungstechnik



ProSafeX
Process Safety Excellence

## Why combine HAZOP and ML based Operational Optimisation ?

atlan-tec
OPTIMIZE YOUR OPTIONS.

**Operational Optimization with Machine Learning is now practical using neural networks in a grey box approach**

- **Development over 2 decades, benefitting from growth in computer power**

**Operational Optimization by Machine Learning can offer big savings, but requires a major effort to provide the data & documents.**
**It must be based on up-to date documentation**

- **Documentation update is a part of HAZOP preparation as well**

**Operational Optimization must stay within safe boundaries**

- **Boundaries can be unclear or overly conservative, if HAZOP not up-to date**

**Operational Optimization can stabilize processes and increase safety level**

**➔ An operationally optimized plant is a safer plant !**

ProSafeX
Process Safety Excellence

## Why combine HAZOP and ML based Operational Optimisation ?

atlan-tec
OPTIMIZE YOUR OPTIONS.

**HAZOPs are usually hated in Operations, but necessary to stay safe**

- **Time & Money saved, when combined with optimization**

**HAZOP is also the deepest analysis of an operating plant after startup**

- **Can't we use this deep analysis also to up performance ?**

**HAZOP is traditionally based on design data.**

- **Shouldn't we include actual operation in a more systematic way ?**

**Suboptimal operation drags down Economic Performance AND Safety**

- **Shouldn't operational optimization also improve Process Safety ?**

**→ This presentations explores a systematic way to make plants safer and more profitable by combining HAZOP and ML based operational optimization**

Dr. Hans V. Schwarz, ProsafeX, 2020 5

ProSafeX
Process Safety Excellence

## Operational Optimisation by Machine Learning

atlan-tec
OPTIMIZE YOUR OPTIONS.

**Building a model which can map process variables to cost per ton**

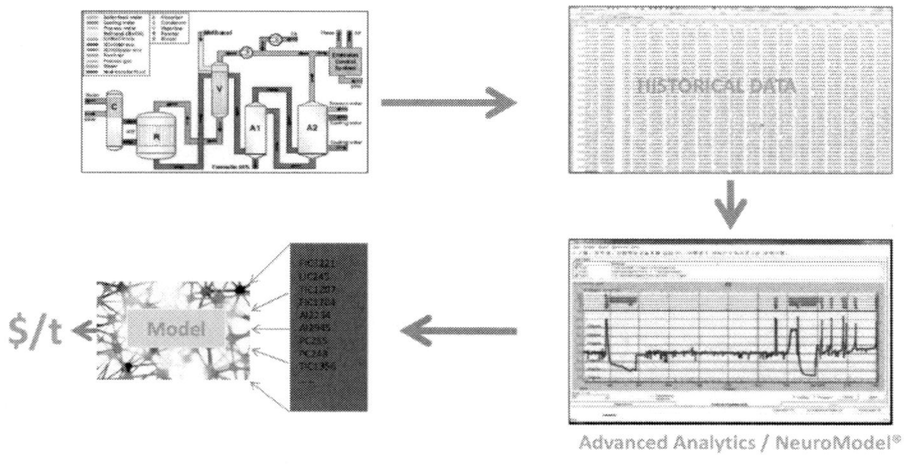


Advanced Analytics / NeuroModel®

Dr. H 6

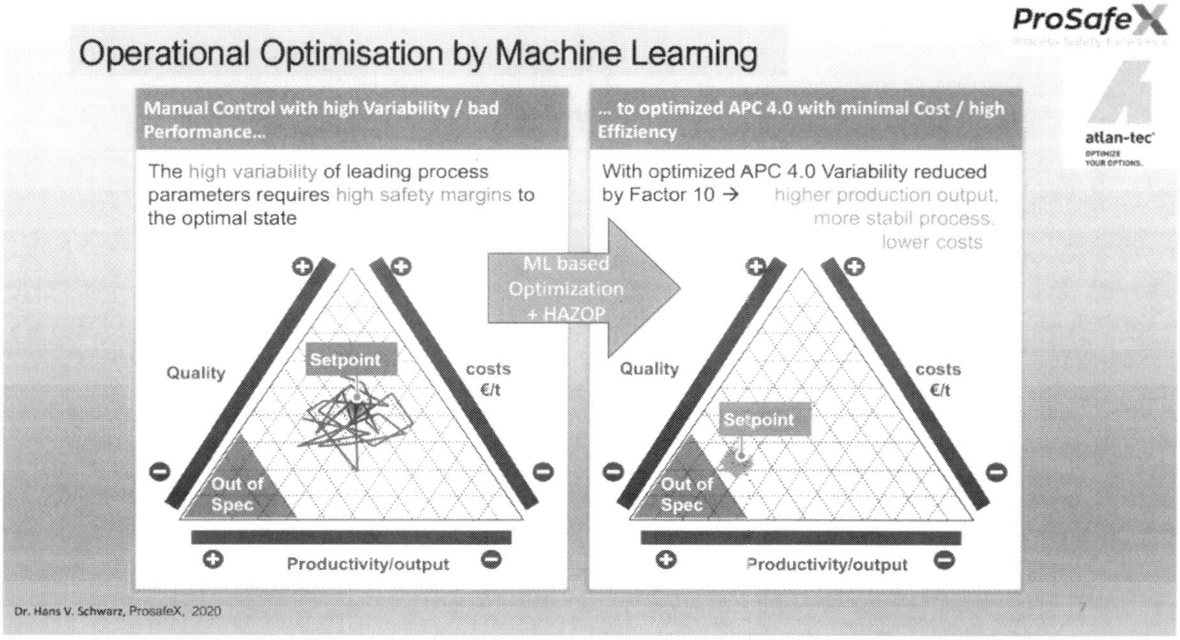
ProSafeX
Operational Optimisation by Machine Learning
atlan-tec
Manual Control with high Variability / bad Performance...
The high variability of leading process parameters requires high safety margins to the optimal state
... to optimized APC 4.0 with minimal Cost / high Effiziency
With optimized APC 4.0 Variability reduced by Factor 10 → higher production output, more stabil process, lower costs
ML based Optimization + HAZOP
Quality
costs €/t
Setpoint
Out of Spec
Productivity/output
Dr. Hans V. Schwarz, ProsafeX, 2020

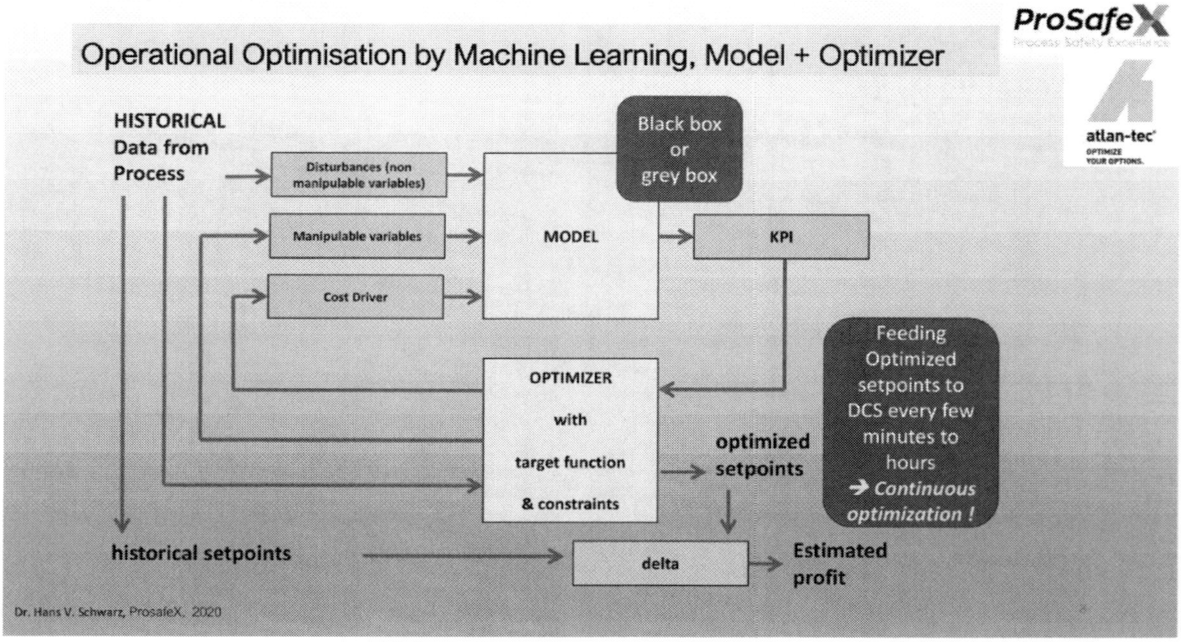
ProSafeX
Operational Optimisation by Machine Learning, Model + Optimizer
atlan-tec
HISTORICAL Data from Process
Disturbances (non manipulable variables)
Manipulable variables
Cost Driver
Black box or grey box
MODEL
KPI
OPTIMIZER with target function & constraints
optimized setpoints
Feeding Optimized setpoints to DCS every few minutes to hours → Continuous optimization !
historical setpoints
delta
Estimated profit
Dr. Hans V. Schwarz, ProsafeX, 2020

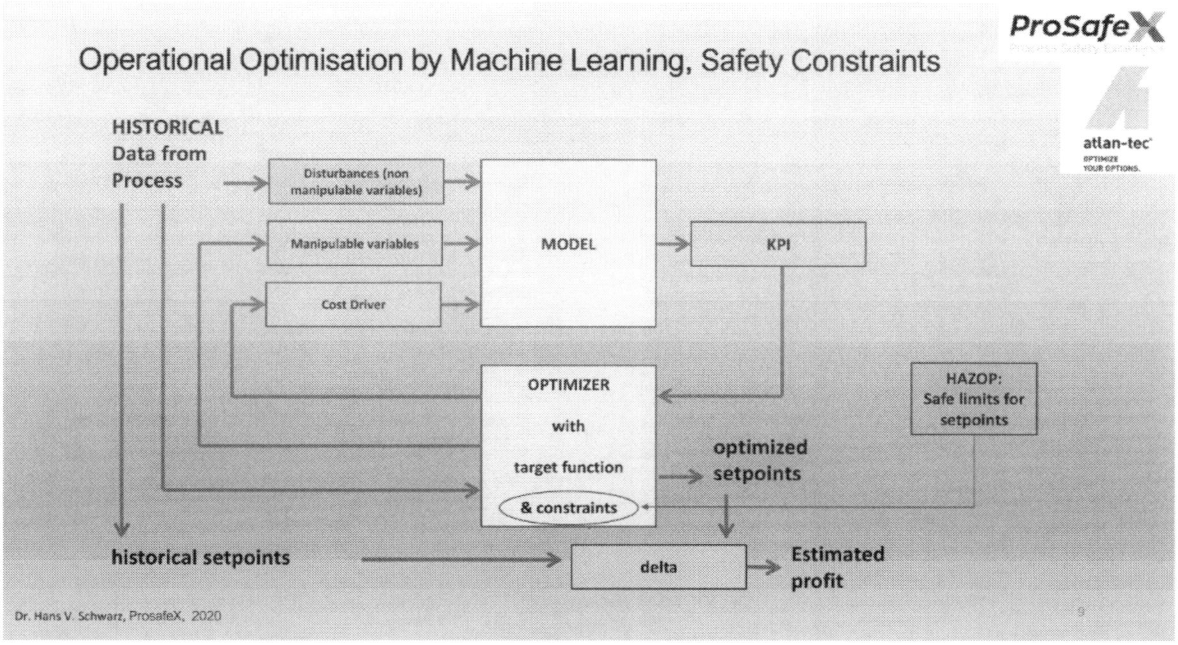


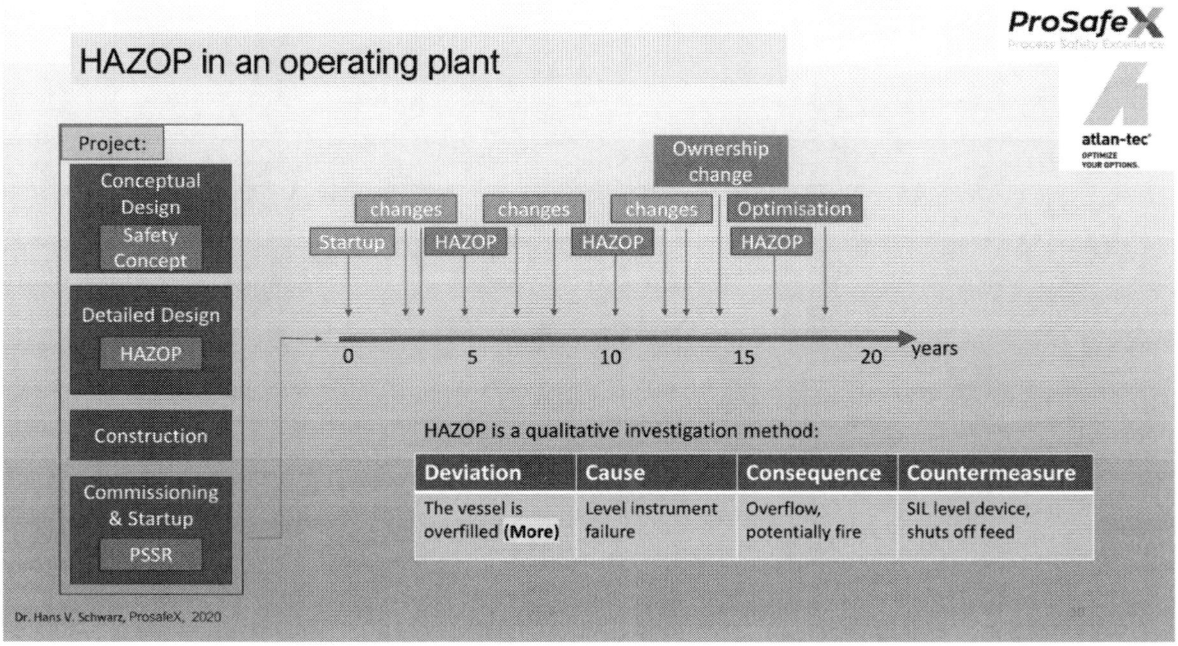


| Deviation | Cause | Consequence | Countermeasure |
|---|---|---|---|
| The vessel is overfilled **(More)** | Level instrument failure | Overflow, potentially fire | SIL level device, shuts off feed |

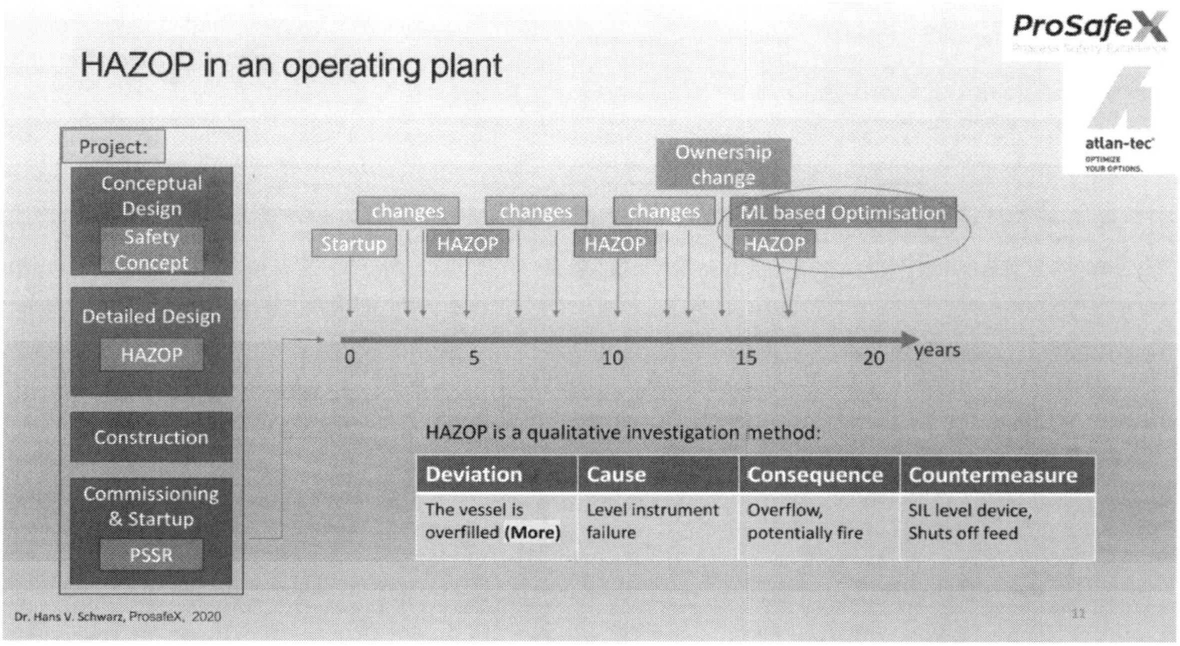


| Deviation | Cause | Consequence | Countermeasure |
|---|---|---|---|
| The vessel is overfilled **(More)** | Level instrument failure | Overflow, potentially fire | SIL level device, Shuts off feed |

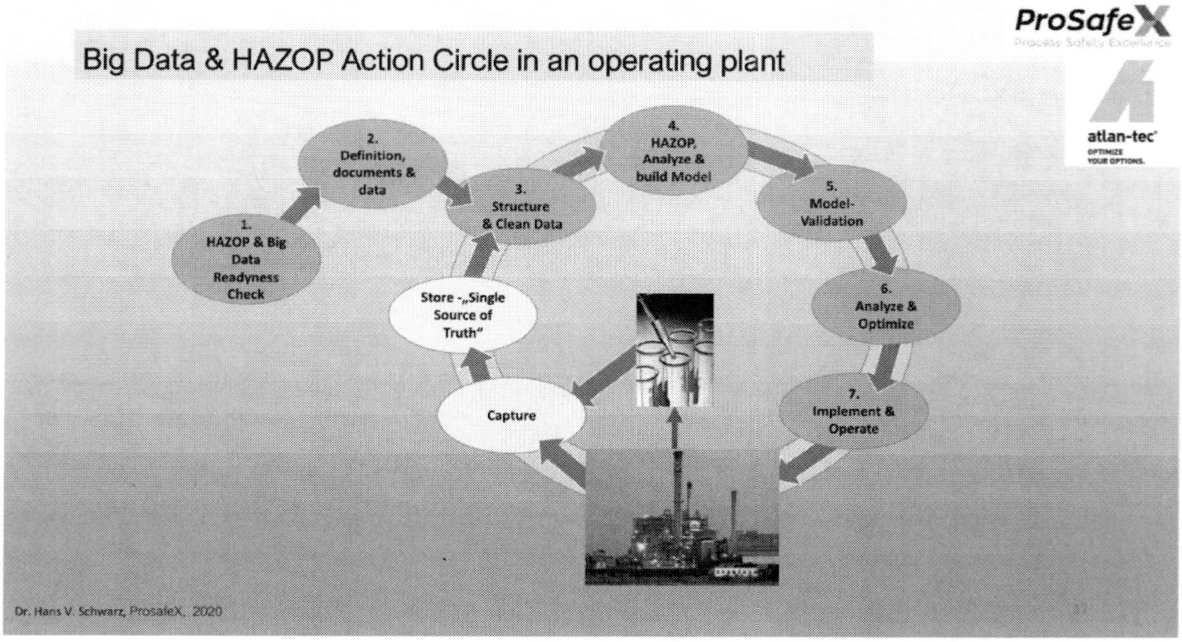

## Synergy in Preparation of HAZOP + Optimisation

**HAZOP as well as Operational Optimisation by Machine Learning require data and documents**

- **P&IDs (Pipe & Instrumentation Diagrams)**
- **Control scheme, functional plans of safety devices**
- **Equipment data**
- **Actual operation history of plant parameters (T, P, F, L,Q,..)**
- **Incidents and deviations**

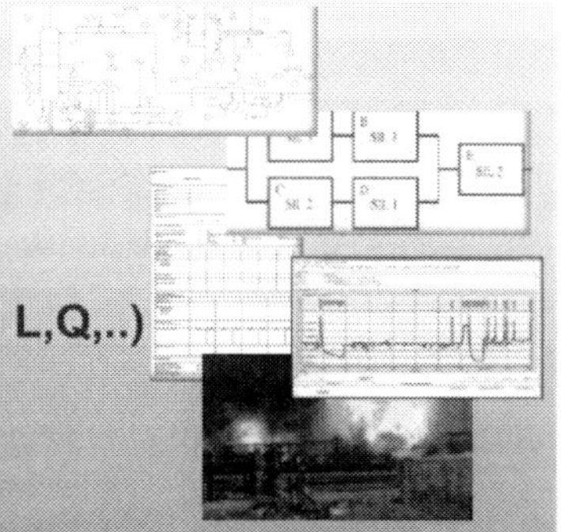

**➔ Big overlap of data needed for HAZOP and for ML based Optimisation**

Dr. Hans V. Schwarz, ProsafeX, 2020 13

## HAZOP and Optimisation ‚Readiness Assessments'

**Preparing the data sounds trivial, but with older plants it is often difficult to have the data ready to ensure a smooth HAZOP or Optimisation study**

- **Readiness Assessments facilitate and guide the preparation**
  - **Atlan-tec's Readiness Assessments for data based optimization**
    - **http://ora.readiness-assessment.com/**
  - **ProsafeX's HAZOP Readiness Assessment**
    - https://smart.atlan-tec.com/415188?lang=en
    - https://www.prosafex.com/services/

Dr. Hans V. Schwarz, ProsafeX, 2020 14

ProSafeX
Process Safety Excellence

atlan-tec
OPTIMIZE YOUR OPTIONS.

## Synergies: HAZOP Quality, Optimisation Risk

**Besides the shared data preparation, there are 2 additional synergies:**

1. **Better HAZOP quality**
   - **As HAZOP is mostly based on design data, it tends to neglect operational experience**
   - **Identified operational weaknesses from historian analysis in the operational optimization study add quality to the HAZOP**
2. **Reduced Risk with Operational Optimisation**
   - **Optimisation must stay within safe parameter boundaries**
   - **The update of the HAZOP study provides fresh input, with a focus on well chosen safety limits – conservative, but not to the extreme**



ProSafeX
Process Safety Excellence

atlan-tec
OPTIMIZE YOUR OPTIONS.

## Summary

- **Synergies of HAZOP and ML based Operational Optimisation**
  - **Shared preparation of data & documents**
  - **Better HAZOP quality by using data from the optimization study**
  - **Reduced risk in plant optimization with pinpointed constraints, through a fresh HAZOP study**
- **Online tools improve efficiency of the process**
  - **Readiness Assessments guide the preparation phase**

Optimized threshold limits of Safety Interlocks and PSVs
Plant runs in optimal operation parameter window
Less unplanned downtime
Fewer safety incidents
Higher yield
HAZOP measures address operational issues
Higher Ebit
Reduced utility consumption

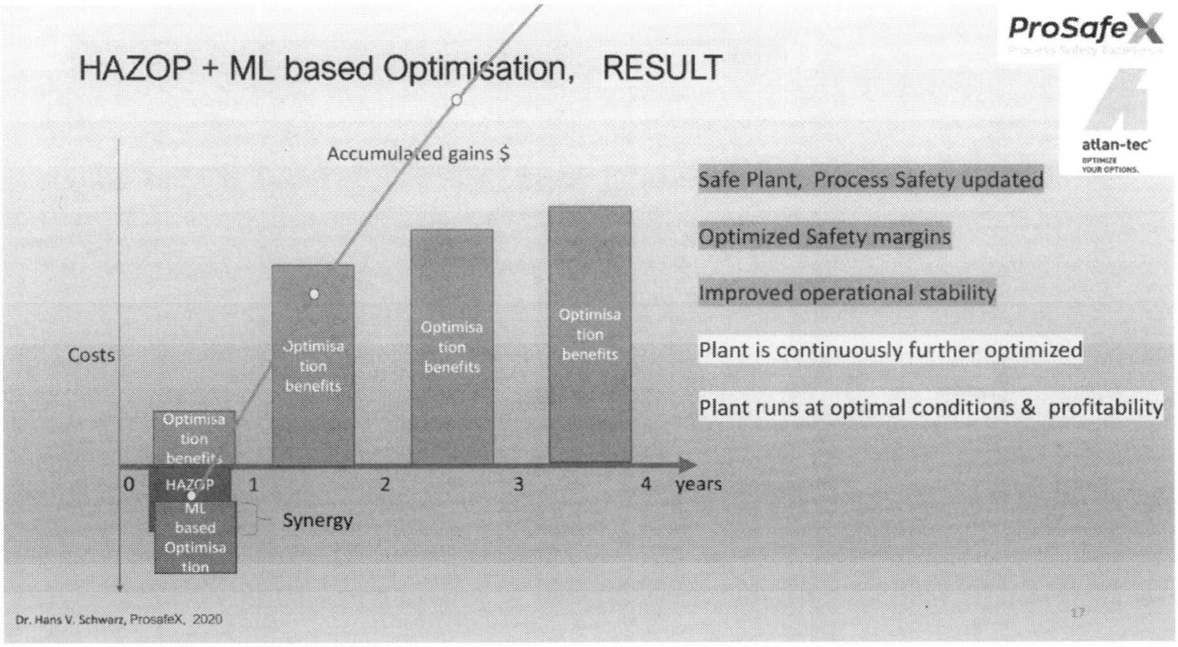


Thank you for your attention !

ProSafeX
Process Safety Excellence

atlan-tec
OPTIMIZE
YOUR OPTIONS.

HAZOP + Optimisation:

Backups

ProSafeX
Process Safety Excellence

atlan-tec®
OPTIMIZE
YOUR OPTIONS.

## Disclaimer

***The recommendations given in this presentation are carefully derived from practical experience in leading chemical companies.***

***However, no guarantees are given that following the advice will always prevent undesired events.***

***The application of the presented information always requires the special situation to be taken into account by the organization and individuals responsible for operation of the asset.***

***No legal responsibility is taken by ProsafeX, Atlan-tec or the presenters.***

Dr. Hans V. Schwarz, ProsafeX, 2020

# HAZOP, in the lifecycle of a plant

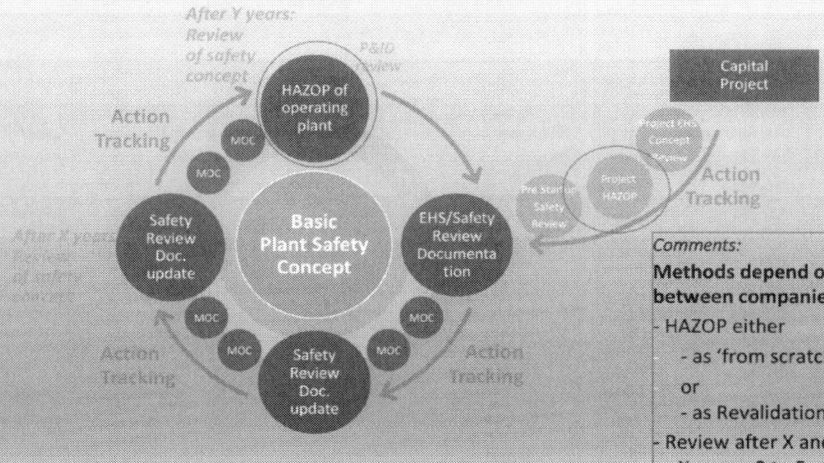


*Comments:*

**Methods depend on Hazard Potential and vary between companies**

- HAZOP either
  - as 'from scratch', new study
  - or
  - as Revalidation Review of prior HAZOP
- Review after X and Y years:
  - X: every 2 to 5 yrs, in some companies
  - Y: every 5 to 10 years, in most majors

Dr. Hans V. Schwarz, ProsafeX, 7/2019

Case 2

# Multi Stage Reactor

**Description**

- This typical example shows a multi-stage process that converts raw materials into a product
- Unknown fluctuations in the raw material causes quality variations
- The ANN model, trained with example data, predicts the Yield and Quality of the process and the GenOpt Real Time Optimizer
- Caused by the prediction, reaction time will be reduced
- Process stability increases by factor 10
- Economical impact: 5% savings

**Block Diagram**

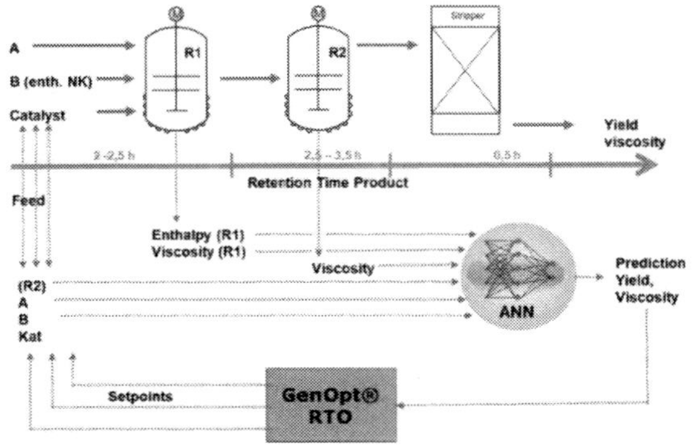


SOURCE: atlan-tec

Case 4

# Methanol plant – Near East

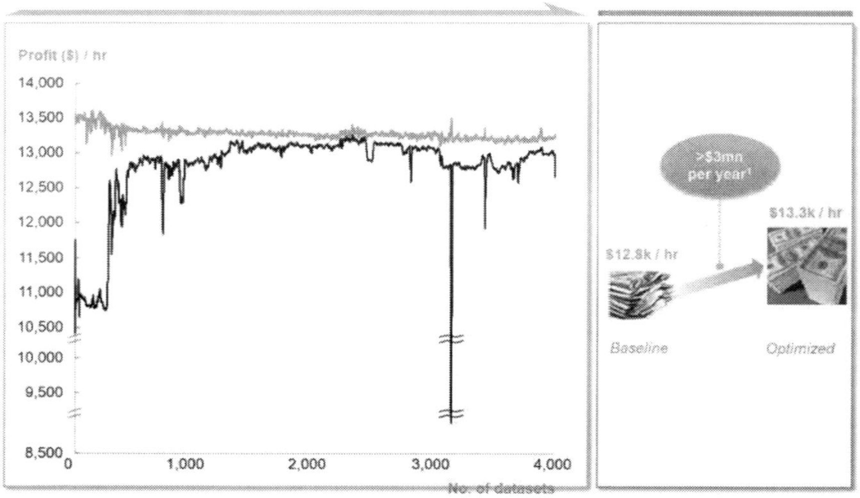


1 Assuming 330 days of continuous production

SOURCE: QUAFAC & atlan-tec

Case 5

# Batch Process Optimization

**Issue in Batch Data: Too many time series and rare quality checks**

**Approach: Generate unique fingerprint per batch to compare**

**Perform Modeling & Optimization using Fingerprints**

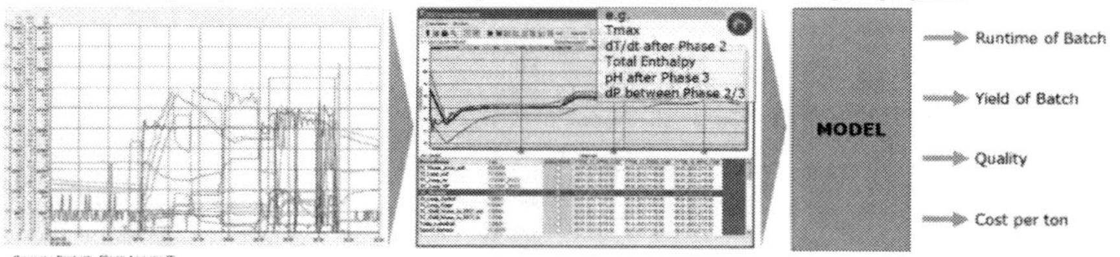


**Some selected Project Results**

- Batch polymerization plant with high variation of run time: Reduction of the average running time from 8.3 hours to 7.2 hours increased use of equipment and higher throughput
- Adhesive production: Reduction of faulty batches from 8% to 3%
- Carboxylic acid / fermentation: Reduction of the frequency of post-fermentation from an average of 3 hours per batch to 0.1 hours per batch
- Organic pigments: Improved yield from an average of 73% to 81% (theoretical maximum is 83%)

SOURCE: atlan-tec

2020 EUROPEAN CONFERENCE ON PROCESS SAFETY & BIG DATA
October 13 – 14, 2020 CCPS EPSC

VIRTUAL

# HSE Management in the Age of Artificial Intelligence. A Brightsite program

Brightsite
Transforming industry

Johan van Middelaar, Sadegh Shahmohammadi (TNO)
Esta de Goede, John van den Hurk (Sitech Services)
Gui Hoedemakers (AnQore)

2020 EUROPEAN CONFERENCE ON PROCESS SAFETY & BIG DATA
October 13 – 14, 2020 CCPS EPSC

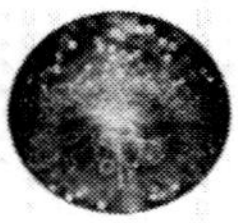

VIRTUAL

- Partners in Brightsite
- HSE Challenges in the chemical industry
- Towards the next era of HSE management
- HSE management powered by AI
- Examples
- Model development and the way ahead

# Partners in Brightsite - working together in living lab Chemelot

# HSE Challenges in the chemical industry

- Inadequate management system performance
- Static conventional risk assessment methods
- Emerging risks related to:
  - New technologies
  - Digitisation (complexity, opacity)
  - .....

## Towards the next era of HSE management

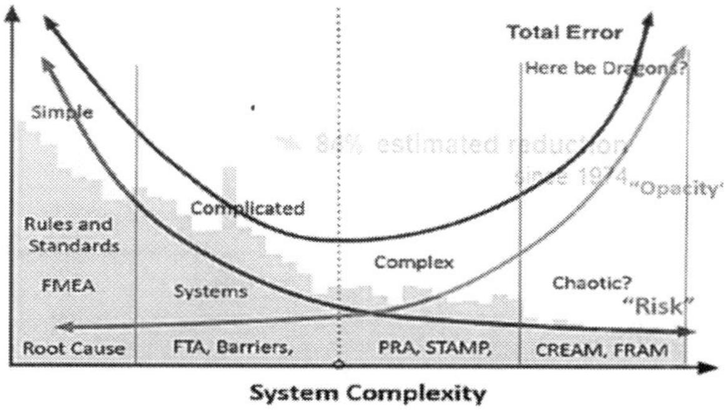


- Enhanced risk assessment of complex systems
- How to identify emerging ('unknown') hazards?

Source: Safety and Complexity, David Slater, July 2020, https://www.researchgate.net/publication/342639150]

## HSE management powered by AI

"Accidents do not occur in a chaotic fashion. Underlying patterns and trends do exist and can be captured."

Source: Sarkar et al. 2019. Computers and Operations Research 106 (2019) 210–224.

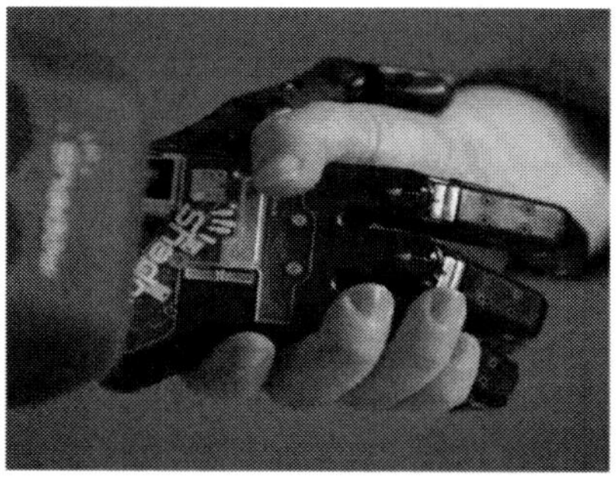

## HSE management powered by AI: Combining various types of data

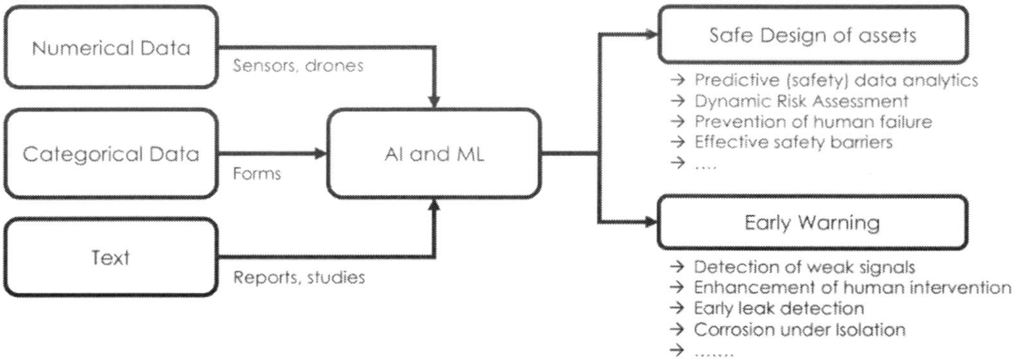


## Example: "Safety Stethoscope": Sentiment Analysis

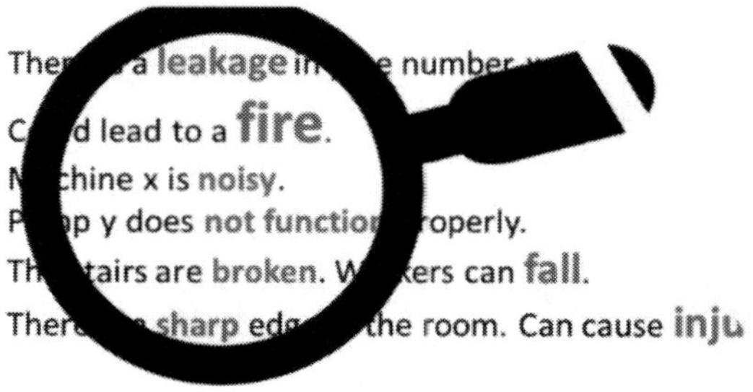


Prevention of incidents by immediate detection of warning reports!

## Example: "Safety Stethoscope": Analyzing Reports

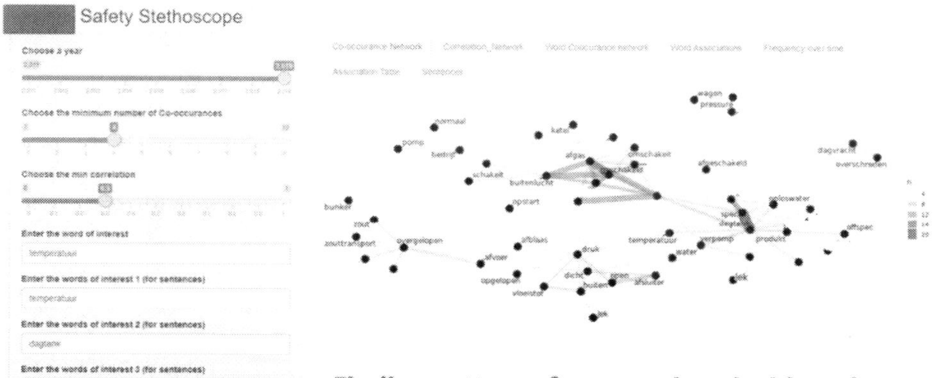


**Finding patterns from previous incidents!**

## Model development: Classification and Prediction of Incidents

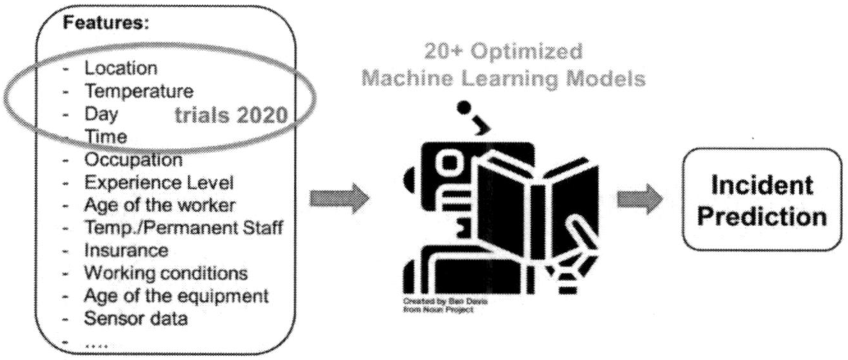

## Model development: HSE Knowledge base

- Intelligent collection, organization and retrieval of knowledge
- Training the machine to learn the safety language

Scenario analyses:

*"What are the risks associated in process x with substance y in condition z?"*

## Model development: Multi-layer Dynamic Early Warning System

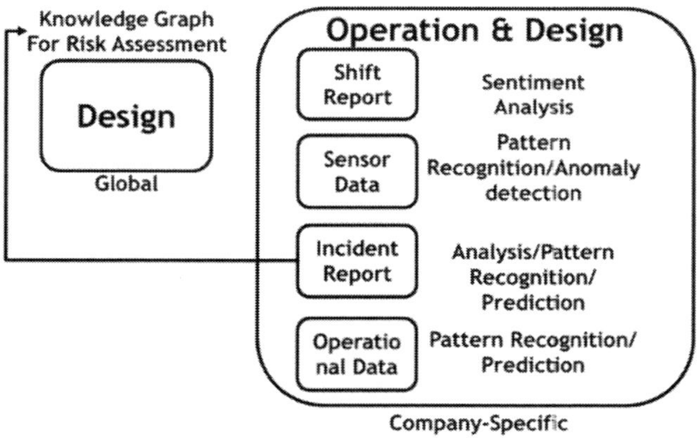

# Model development: Technical Readiness Planning

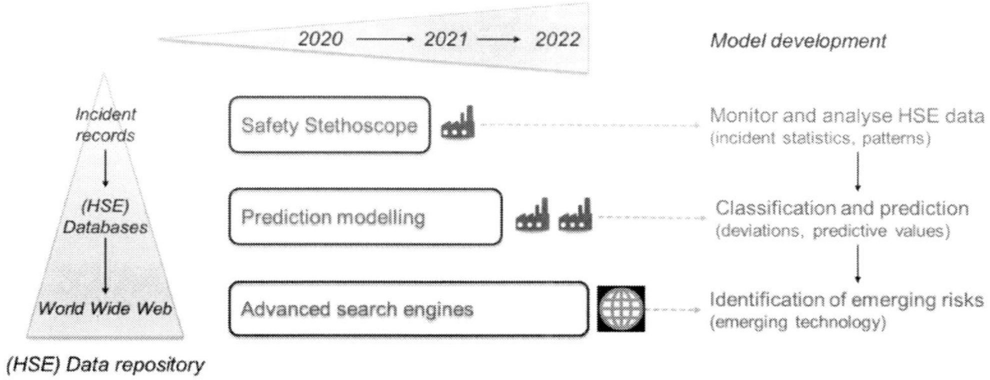


# The power of Brightsite

Model/techniques development, testing in commercial plants and demonstrating added value.

In this use case:

- Advanced AI/ML techniques in the complex operational environment of a chemical plant
- Added value results
  - Predicting safer operations
  - Providing options for pro active actions

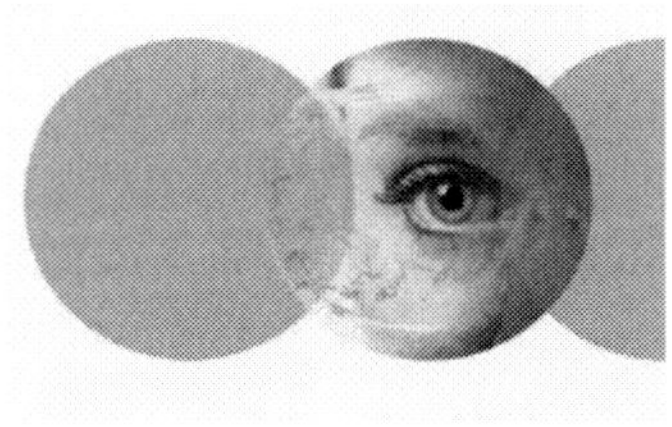

## The way ahead

(Brightsite) partnerships with future partners:

- (Petro) Chemical industry
- (Technical) Universities
- Process Safety Associations
- 'Safety Delta Netherlands'
- you?

2020 EUROPEAN CONFERENCE ON PROCESS SAFETY & BIG DATA

October 13–14, 2020

CCPS EPSC

VIRTUAL

# Thank you !


johan.vanmiddelaar@tno.nl

sadegh.shahmohammadi@tno.nl

esta.goede-de@sitech.nl

john.hurk-van-den@sitech.nl

gui.hoedemakers@anqore.com


Brightsite
Transforming industry

TNO innovation for life

sitech services

AnQore

2020 EUROPEAN CONFERENCE ON PROCESS SAFETY & BIG DATA

October 13 – 14, 2020

CCPS EPSC

VIRTUAL

# Knowledge driven online FMEA system for risk management

TOTAL Refining & Chemicals – Refinery 4.0

Nicolas Ranchet

## TOTAL Refining & Chemicals - Refinery 4.0 program

TATA CONSULTANCY SERVICES

TOTAL

- Safety
- Competitiveness
- Environmental footprint
- Agility

Inventing and developing Digital Innovations, benefiting from experiences across industries

- Artificial Intelligence, Data Science
- Internet of Things, Edge computing
- Automation, Interoperability
- Optimization, decentralised decisions

To transform our way to operate, supervise, inspect, maintain, optimize the refinery.

TOTAL Refining & Chemicals – CIDI Refinery 4.0 – 2020 European conference on process safety & big data – October 13th

TOTAL

## FAILURE SCENARIOS DETECTION

Many risk analysis documents contain the expertise and lessons learned of the subject matter experts and the field teams. New or improved analysis are generated on a regular basis.

The information is stored, under different format (structured or unstructured), and located in various information environment.

The knowledge stored in those documents has to be available to the operation team in the right context of the processing unit operation. So that the operation team takes the best action considering relevant insights.

Two solution components have been developed

→ Insight engine

→ Online FMEA engine



## INSIGHT ENGINE – PRINCIPLES & VISUALS

The insight engine extracts and presents failure related information from documents

root cause analyses
Process Hazard Analysis
incident reports
return of experience
maintenance notifications

SINEQUA

thesaurus
- failure modes
- causes
- effects

family of Equipment

equipment to functional location match

failure – causes – effects relations

contextual document list

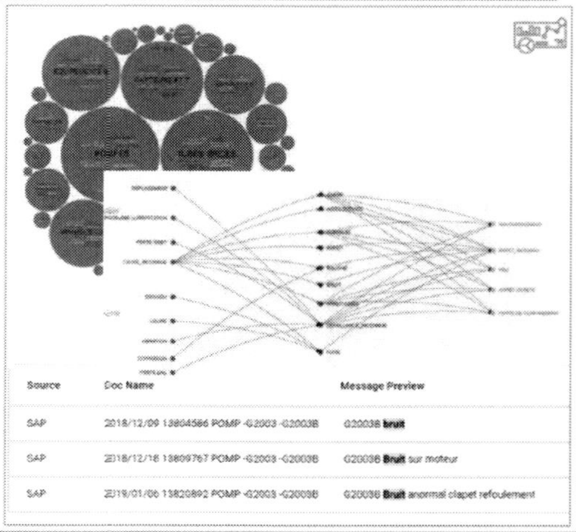

## ONLINE FMEA ENGINE - PRINCIPLES

Main objective : to identify current and possible failures and to issue alerts to the operator.

TCS PREMAP

incident analysis
HAZOP analysis

→ knowledge repository → knowledge driven reasoning engine → failure path identification →

real time process values alarms ↑ (into knowledge driven reasoning engine)

workflow periodic execution

Gautham et al, 2013

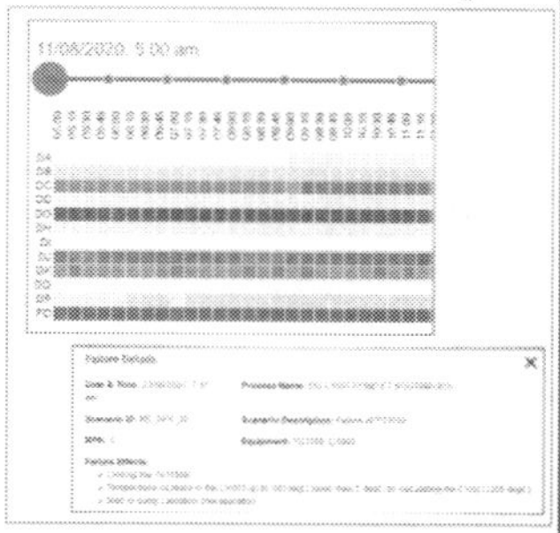


offline knowledge curation & modeling

online risk analysis

TOTAL Refining & Chemicals – CIDI Refinery 4.0 – 2020 European conference on process safety & big data – October 13th

TOTAL

## ONLINE FMEA ENGINE – FROM KNOWLEDGE TO DATA

The knowledge repository is a collection of **knowledge elements** defining diverse failure paths able to

- detect a failure mode and
- trigger the appropriate effect

(Reddy et al., 2017)

| Failure mode detection | ↔ | Effect trigger |
|---|---|---|
| mathematical conditions on process parameters and alarms | | effect conditions on process parameters and alarms |

(FCC:FC2080.pv < 30.00) &&
(FCC:C2G1027.pv == ARRET ||
FCC:652G1009A.pv == ARRET &&
FCC:652G1009B.pv == ARRET)

Loss of discharging C1003 : FCC:FC2080.pv < 30.00
Increase in the level C1003 : FCC:C2LC2069.pv > 70.00
Temperature increase above the tray 10 : FCC:C2TI2045.pv > 179.99

An effect can be a subsequent failure mode following the 'failure mode effect ontology' when matching the right context.

It leads to identify a chain of failure events and effects starting from a root cause

TOTAL Refining & Chemicals – CIDI Refinery 4.0 – 2020 European conference on process safety & big data – October 13th

TOTAL

## ONLINE FMEA ENGINE – FROM KNOWLEDGE TO DATA

In the context of online FMEA, a knowledge element is defined through two ontologies
- failure mode and effect ontology
- equipment or process system ontology

failure mode and effect ontology

| Reference document name | Entry index | Detection parameter | Failure Mode | Impact | Impact-detection |
|---|---|---|---|---|---|
| Hazop Node1, DS.2 | 51 | Reduction of cumulative flow in FT2036 A+B+C+D OR Pump status from DCS (FCC:852G1001A.pv == 0 && FCC:852G1001B.pv == 0) | | | |
| | | | Stopping the G1001 | | |
| | | | | Reduction of throughput at the pumps G1001 until max flow of J595/J596 (either 180 t/h) see less flow | D(FT2036 A+B+C+D)/dt <<0 (sharp decrese) |
| | | | | Pressure drop in the node | PC2072.pv < 4.8bar |
| | | | | Without consequence security on this node | |

equipment ontology

| Process | Measured parameters | Sensor | Equipment |
|---|---|---|---|
| | | | |
| Charge et ligne de transfert | | | |
| | Flow rate | FT2036 A, B, C, D | G1001 |
| | Pressure | PC2072 | |
| | | | |

In this example, the knowledge is extracted from the FCC PHA. It describes a pump stoppage failure mode and its effects.
The equipment ontology defines some relationships between the equipment. E.g : FT2036 is sensing the G1001 discharge flow rate



## ONLINE FMEA ENGINE – SOLUTION VALIDATION

The online FMEA solution has been validated against a real-life incident in the FCC unit of an oil refinery : sensors data stream of an actual incident is passed through the Online FMEA engine, the failure related to compressor stoppage is indicated in the output.

1. The failure mode suggested in the detailed incident analysis matches with the failure detected by the online FMEA solution
2. The incident analysis report refers to observed effects, which match with those predicted by the ontological context
3. The detection and report generation by online FMEA solution takes under 5 minutes.

## ONLINE FMEA ENGINE - VISUAL

## MAIN CHALLENGES AND NEXT STEPS

Populating and refining the domain knowledge model

- Both knowledge modelling and process engineering skills are required
- augmentation of the knowledge model content requires a continuous improvement loop

Modelling operator actions from past experience and simulation
Integrating remedial actions along with knowledge elements

Thank you for your attention

TOTAL

PREVENTIVE MAINTENANCE
& MACHINE LEARNING
PREVENTIVE MAINTENANCE BY MAKING USE OF MACHINE LEARNING
AND CURRENTLY AVAILABLE INSTRUMENTATION
Bart Groenenboom - Reliability Engineer - DSM Food Specialties
14-10-2020
DIGI
DICI GOES DIGITAL
DSM

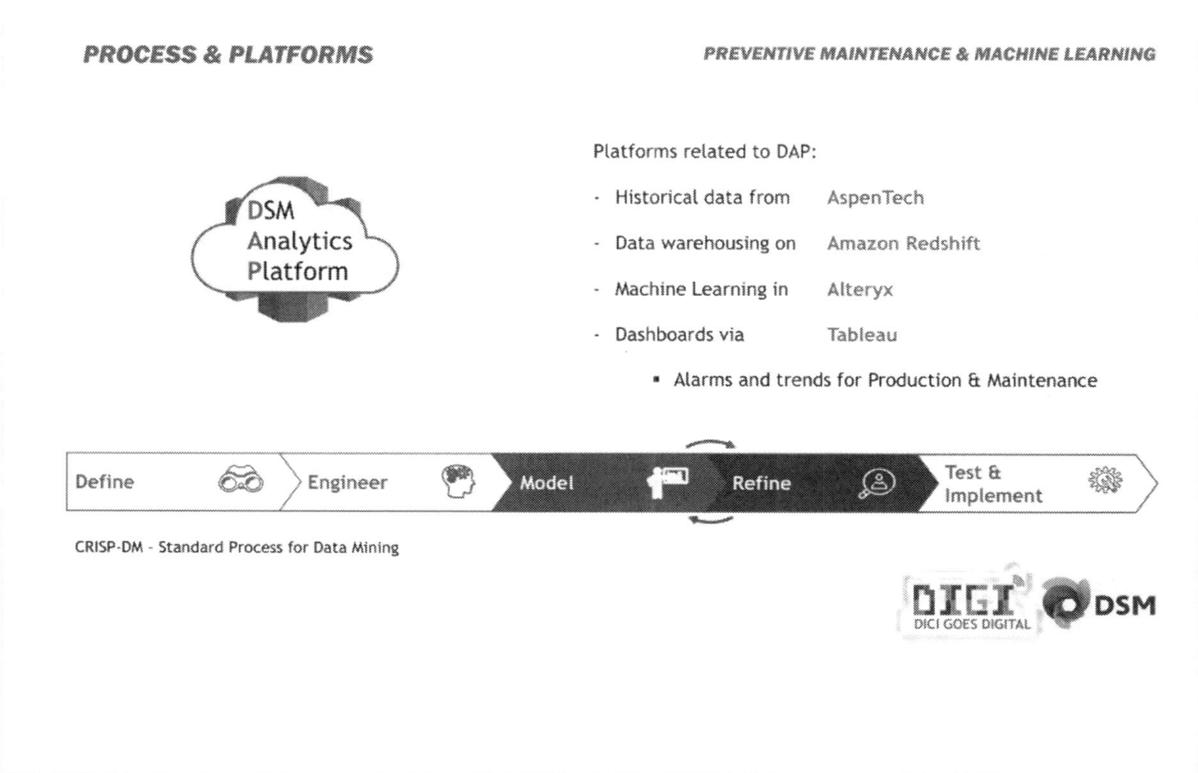
PROCESS & PLATFORMS
PREVENTIVE MAINTENANCE & MACHINE LEARNING
DSM
Analytics
Platform
Platforms related to DAP:
- Historical data from AspenTech
- Data warehousing on Amazon Redshift
- Machine Learning in Alteryx
- Dashboards via Tableau
▪ Alarms and trends for Production & Maintenance
Define
Engineer
Model
Refine
Test & Implement
CRISP-DM - Standard Process for Data Mining
DIGI
DICI GOES DIGITAL
DSM

**DATA UNDERSTANDING** | **PREVENTIVE MAINTENANCE & MACHINE LEARNING**

Define

Table columns known

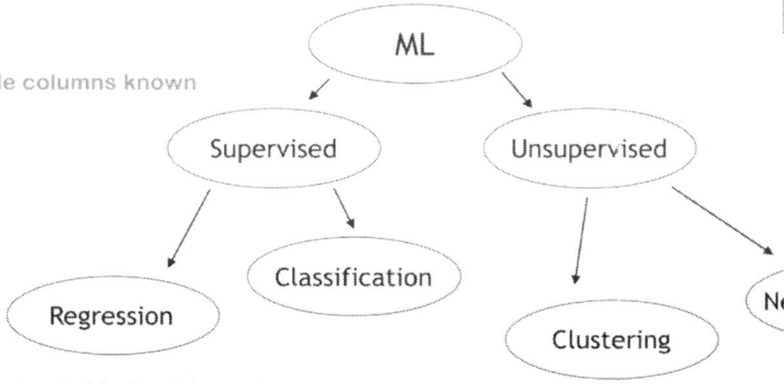


Most suitable for time-series

- Multivariate Linear Regression
- Cox Regression (survival analysis)

DIGI DICI GOES DIGITAL DSM

## MULTIVARIATE LINEAR REGRESSION
## -- YEAST CENTRIFUGES --

DIGI DICI GOES DIGITAL DSM

Engineer

Model

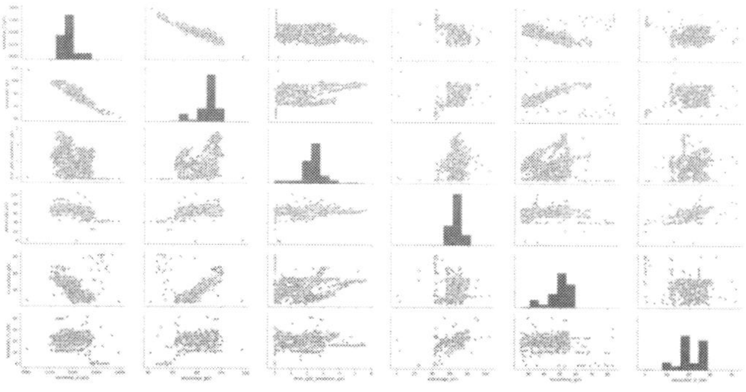

$$Y = B_1X_1 + B_2X_2 + B_3X_3 + \dots$$

Y = Current G01

$X_1$ = Revolutions
$X_2$ = Current P01
$X_3$ = Flow feed G01
$X_4$ = Flow water
$X_5$ = Pressure water

DIGI DICI GOES DIGITAL DSM

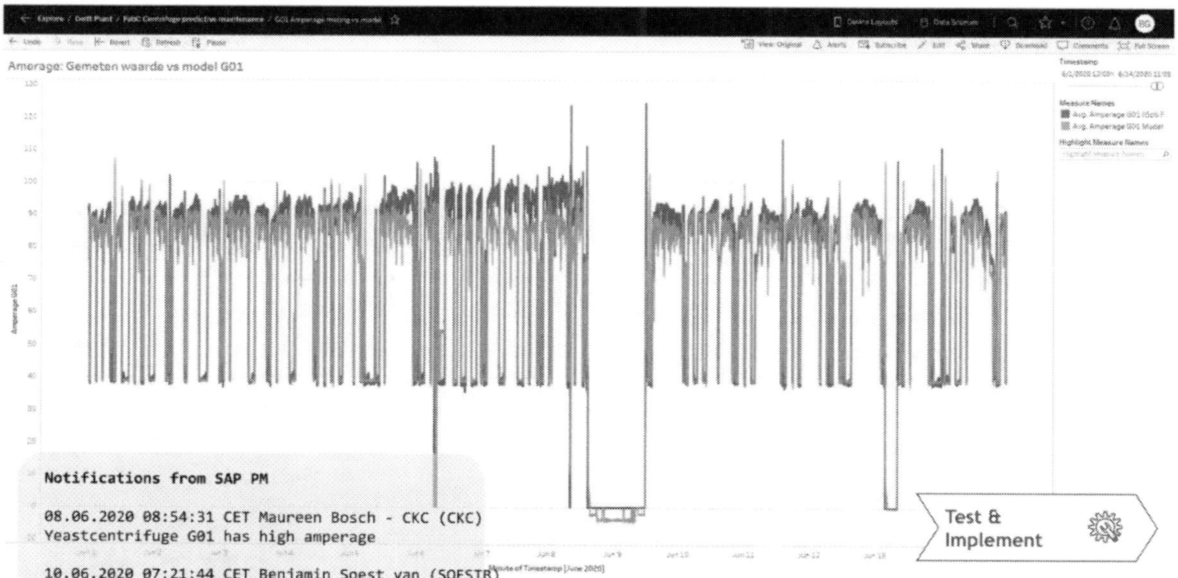


**Notifications from SAP PM**

08.06.2020 08:54:31 CET Maureen Bosch - CKC (CKC)
Yeastcentrifuge G01 has high amperage

10.06.2020 07:21:44 CET Benjamin Soest van (SOESTB)
Centrifuge opened and found worn out sealcups

Test & Implement

DIGI DICI GOES DIGITAL DSM

## *SURVIVAL ANALYSIS WITH PROPORTIONAL HAZARDS*
## *-- YEAST CENTRIFUGES --*

DIGI DICI GOES DIGITAL DSM

## *SURVIVAL ANALYSIS USED FOR MEDICINE*

## *PREVENTIVE MAINTENANCE & MACHINE LEARNING*

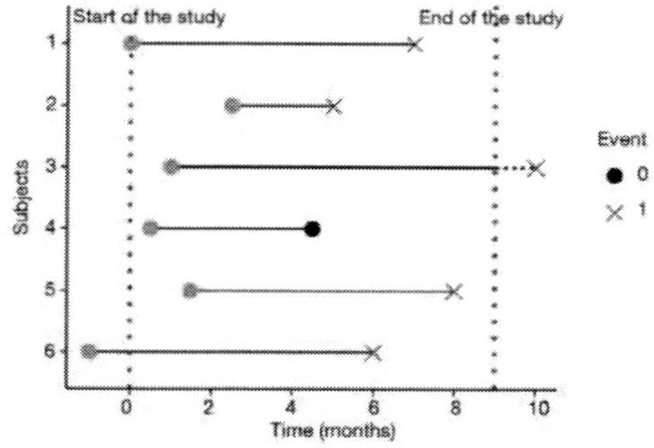


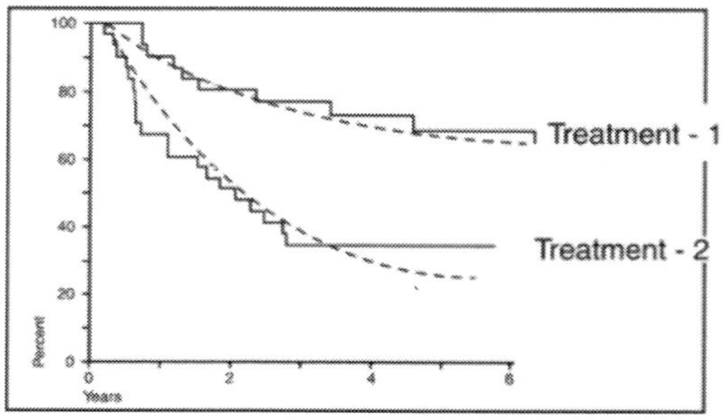


DIGI DICI GOES DIGITAL DSM

## SURVIVAL ANALYSIS FOR CENTRIFUGES

**PREVENTIVE MAINTENANCE & MACHINE LEARNING**

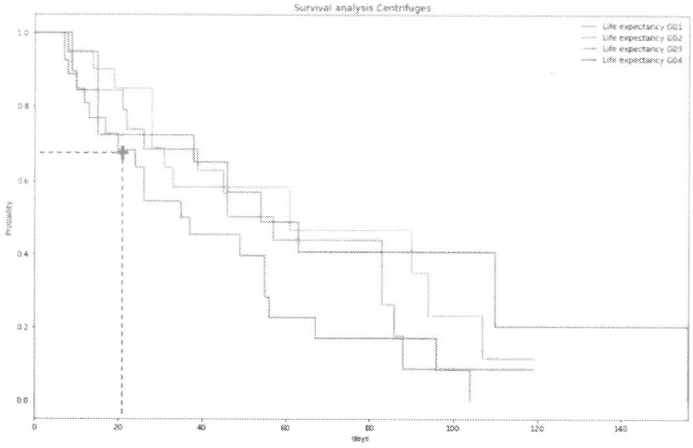


Survival analysis for Yeast Centrifuges:

- 162 failure of all centrifuges

Example:

- G01 has a 67% probability of 3 weeks running without failure.

## SURVIVAL ANALYSIS WITH HAZARDS

**PREVENTIVE MAINTENANCE & MACHINE LEARNING**

Positive influence on lifetime (with spread) →

Little influence on lifetime (with spread) →

Negative influence on lifetime (with spread) →

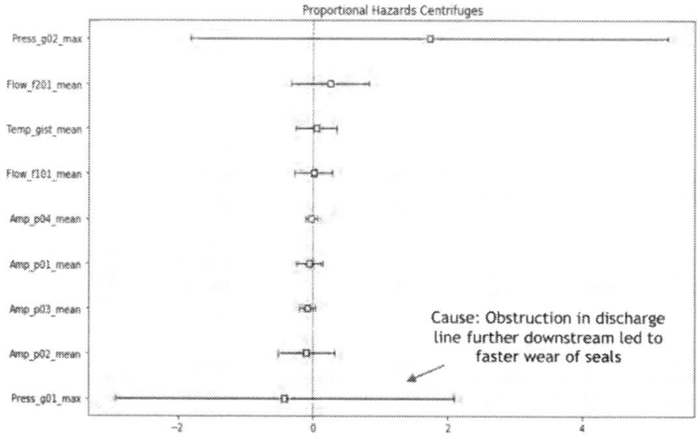

## SURVIVAL ANALYSIS PREDICTED LIFETIME



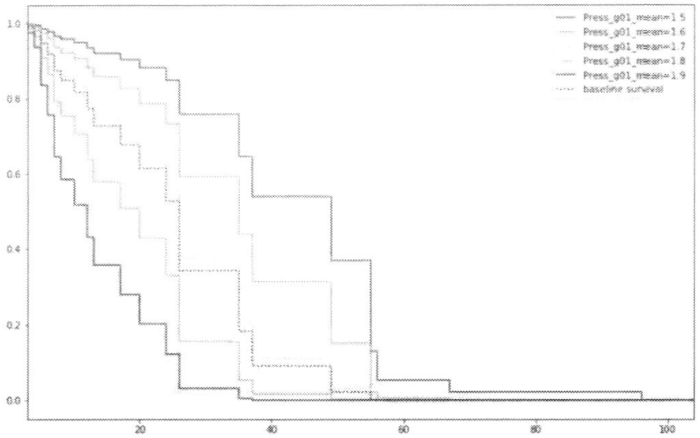


Predict lifetime with different scenarios.

Example: use for adjusting process settings to keep running until next turn around

DIGI DICI GOES DIGITAL DSM

## NEXT STEPS FOR DSM FOOD SPECIALTIES



Implemented regression models:

- Yeast centrifuges (4)
- Steam boilers (4)
- Heat-exchangers in boilers (12)
- Steam pressure control-valves (12)

Future opportunities:

- Air compressors (5)
- Decanters (2)
- Yeast reactors (6)

Survival analysis / Cox regression:

- Survival analysis as part of RCA / FMECA / HAZOP
- Cox regression for predicting maintenance on scenarios

DIGI DICI GOES DIGITAL DSM

## CONDITION BASED MAINTENANCE – MODEL CENTRIFUGES

Slide Platforms and CRISP-DM

One year ago we started with Machine Learning for Condition Monitoring as part of the 'Delft Integral Continuous Improvement' in search for the 'Golden Batch'. Platforms were therefor already up and running. These platforms are AspenTech for process measurements, Amazon Redshift to extract, transfer and load them (ETL). Alteryx reads data from the server, calculates new values and writes them to the same server. Tableau is a business intelligence tool to read and display the data from the server. It also has the possibility to send alerts by mail to it's users which is great for the purpose of predictive maintenance. No need to look at the trends all day. Just an alert when equipment starts to show certain behavior.

A general approach for data analysis is the Cross Industry Standard Process for Data Mining. I will take you through the steps that we've taken at DSM to come models for predictive maintenance.

Slide Machine Learning methods

- Maybe this is a re-run of other presentations today but I would like to give you some short insight in the methods in Machine Learning. First of all there are two big difference in the big-data and Machine Learning. It can be supervised where the data has known columns as you have with data coming from Aspen. Then there is unsupervised learning, for example clustering of internet users with the same behavior. It is unknown what this behavior precisely is. Just that user appear to look alike.
- For the supervised learning you could use classification where you for example predict a risk category (high/medium/low) or you could use a regression analysis where you predict a value, for example a temperature (in case of weather prediction) or a valve opening in case of the industry. For time series, the regression analysis is most suitable. I will show you our examples two types of regression, Multivariate Linear Regression and Survival analysis (or Cox Regression).

DIGI DICI GOES DIGITAL DSM

## CONDITION BASED MAINTENANCE – MODEL CENTRIFUGES

- Together with our other Reliability Engineer, Gaitrie Kalloe, we started with our bad actor the four Yeast Centrifuges. Bad actors, however with not so many comparable failures. As a result, training a classification algorithms (e.g. high/medium/low risk) based on the failures was hardly possible.

- Instead of looking how close you are to failure, it is easier to see how far you are from 'normal' operating conditions. A regression algorithm is an easy was to achieve this. The yeast centrifuges are equipped with measurements of the current and vibrations which are a perfect indicator for failure modes like imbalance, internal leakage and contamination.

Slide Correlations

- Modelling in Python gives you the possibility to find correlations between multiple and unexpected measurements (or features) which could predict an estimator. You could do this in two ways. Start with a logical features to predict your estimator and start expanding the model by adding features which improve your model (modelling takes time but it saves time for data loading). The other option is to start with many features which are in some way related to the estimator and drop the features which are not adding any accuracy of the model (so called 'pruning'). The modelling is easier but the Data Preparation might be more work and slow as you need to prepare and load all the features.

DIGI DICI GOES DIGITAL DSM

## CONDITION BASED MAINTENANCE – MODEL CENTRIFUGES

- When the model is trained and tested on a period the system was functioning well it can be scheduled to calculate the result of the new day/hour/minute. Then it will give a real-time comparison between the estimated values and the actual values. This gives you the possibility to see deviation right at the point they start to happen.

Slide Tableau

- Depending on the time to failure of the failure mode the right person needs to be notified. In the example of the centrifuge the internal leak start and the centrifuge would fail on over-current in less then a week. In this case we needed to inform the operator to write a maintenance request when a deviation would occur. It would save us several days to plan the work while the centrifuge was still running. With four centrifuges this gave less interruptions of the work schedule of the maintenance department.

- For longer times to failure the Maintenance Engineer and/or the Process Engineer should be informed. For example when a heat exchanger gets contaminated, the plug of a control valve starts to wear or the efficiency of a steam-boiler starts to decrease. In that case the cleaning or replacement can be planned.

- Now when Maintenance is planned and executed the model needs to be retrained again for the new situation. This is a challenge. Replacement of a part is pretty clear, but how about a major overhaul or a small overhaul? This is something we need to figure out. Does the Maintenance Engineer need a reset button somewhere on the dashboard? Or can we automate this by using SAP work orders? As we don't have many models yet we can still do this manually, but when we scale up this needs attention.

- This was one example of the regression model we have set up. Now I'd like to give an example of a survival analysis we have done.

DIGI DICI GOES DIGITAL DSM

## CONDITION BASED MAINTENANCE – MODEL CENTRIFUGES

Slide Survival analysis used for Medicine

- As the theory behind most algorithms are over more then a 100 years old, there are several fields of expertise where they've been used in for many years. We can benefit a great deal from these fields if we see the parallels to maintenance. An example is the survival analysis. This analysis is used in medical research to estimate life expectancy for people.

- For example, we would like to know the average life expectancy of a man in the Netherlands. A survival analysis can be used for that just by looking at their age at the time of death. However we could also differentiate by estimating the life expectancy of a smoker and a non-smoker. Smoking is called a proportional hazard. To do this we use the Survival Analysis with Proportional Hazards, also the Cox Regression. Now we can see the adjusted life expectancy for different scenario's. For example also different treatments of a patient.

Slide Survival analysis for Centrifuges

- Using the same theory for the life expectancy of centrifuges gives this figure. It gives the expectancies for 4 centrifuges based on all their failures leading to downtime no matter what the failure mode was (similar to the average age of a man in the Netherlands).

- Now we look for proportional hazards for the centrifuge which could have influenced the life time.

Slide Survival analysis with Hazards

- This plot shows the average life expectancy (dotted line) and the influence of the process conditions on the number of failures with the confidence interval. It shows that a higher discharge pressure of G01 has a negative influence on the expected lifetime. And a higher discharge pressure of the G02 has a positive effect. In the end the discharge line of the G01 was blocked giving a higher pressure. Not known at that time was that this pressure apparently has an effect of the seals in the centrifuge causing internal leaks to appear quicker.

DIGI DICI GOES DIGITAL DSM

## CONDITION BASED MAINTENANCE – MODEL CENTRIFUGES

Slide Survival analysis predicted lifetime

- The Cox Regression can now predict the life expectancy for different scenarios for the proportional hazard. These outcomes can be useful to show the effect of higher demand of equipment. Also it enables you to give a base for decreasing your maintenance interval when the proportional hazards are reduced.

Slide Next steps for DFS

- Up to now we have implemented the linear regression models for the centrifuges, steam boilers, heat exchangers and steam control valves. We have studied our bad actors further to see where we see opportunities for models. It is based on the detectability of the main failure modes and the number of useful measurements near the equipment.

- The survival analysis is mainly going to be used for Root Cause Analysis and to estimate the probability of failure as input for risk analysis like FMECA and HAZOP. I believe this is the a big opportunity for process safety studies for aging assets where no accurate failure data is available.

- Thank for your attention.

DIGI DICI GOES DIGITAL DSM

# Asset Health Monitoring

Assessing Mechanical Asset Integrity by Means of Permanently Mounted Sensors

Mehr Wert.
Mehr Vertrauen.

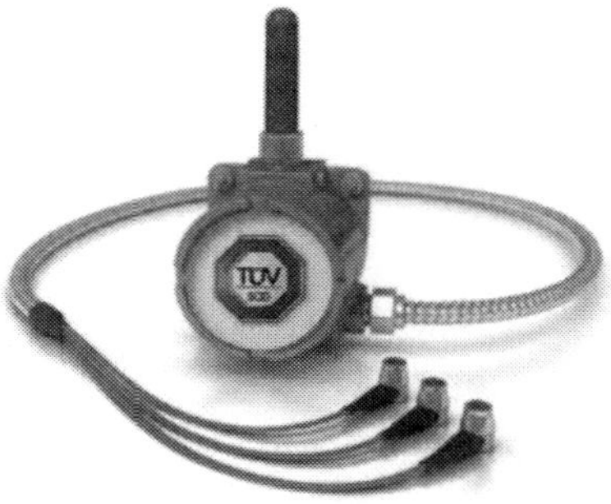

**Presenter:**

Dr. César Buque

Global Manager Mechanical Integrity
TUV SUD Group
Mobile +49 175 9876 365
cesar.buque@tuvsud.com

Presentation to the 2020 European Conference on Process Safety and Big Data
October 13th – 14th, 2020



"Did you know that **Corrosion** costs us an astounding **2.5 Trillion USD** globally?"

-- The World Corrosion Organization

# Content

- The challenge or why ASSET HEALTH MONITORING?
- Current inspection practice and corresponding limitations
- The ASSET HEALTH MONITORING Solution
- ASSET HEALTH MONITORING Approach
- ASSET HEALTH MONITORING Sensors – Example for Wall thickness monitoring
- Value added of ASSET HEALTH MONITORING DATA for Process Safety
- Typical applications in Process Safety
- Summary



# The Challenge or why ASSET HEALTH MONITORING?

“ We keep comply with all **regulatory frameworks.** We perform all **periodical inspections** according to **standards**. Therefore we were all very **surprised by the failure** that led to this **explosion**. ”

BP Senior Speaker
ZDF November 12th, 2004

## Current inspection practice and corresponding limitations – 1/3

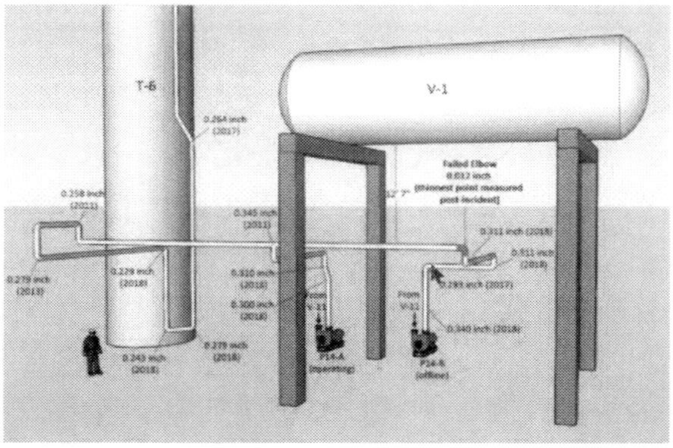


### Current practice

- Plant operators perform spot or scan inspections yearly
- PPPP Effect not taken under consideration
- Conservative design specifications (bigger WT)
- Failures, leakage, unplanned repairs and/or shutdown



## Current inspection practice and corresponding limitations – 2/3

### Disadvantages of current practices

- PPPP-Effect (Position, People, Procedure, Probe)
- Spot measurements have a large error band width, so:
- Not appropriate for prediction of corrosion trends
- Not frequent enough to derive corrosion trends
- Increasing inspection frequency means higher costs
- Failure prediction requires dynamic FFS information

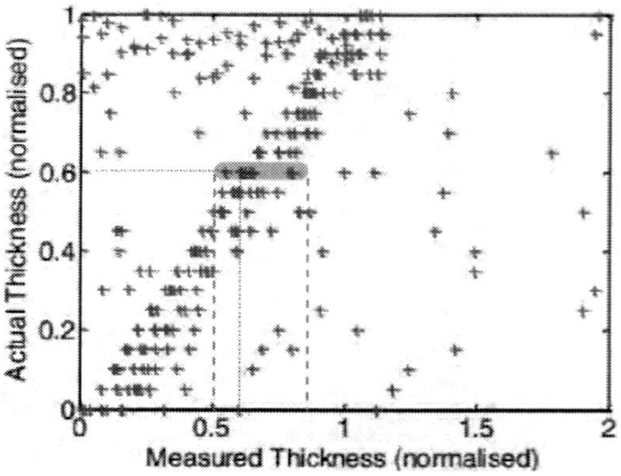


Measurement errors associated with manual ultrasonic thickness measurements on a corroded and retired steel sample

# Current inspection practice and corresponding limitations – 3/3

## Disadvantages of current practices

- PPPP-Effect (Position, People, Procedure, Probe)
- Spot measurements have a large error band width, so:
- Not appropriate for prediction of corrosion trends
- Not frequent enough to derive corrosion trends
- Increasing inspection frequency means higher costs
- Failure prediction requires dynamic FFS information

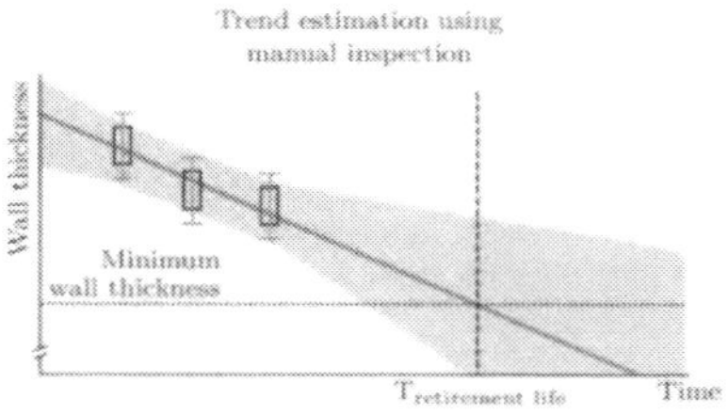


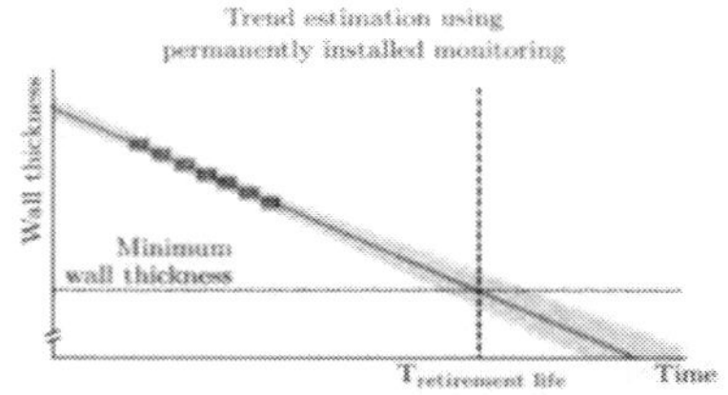




# THE ASSET HEALTH MONITORING APPROACH – 1/2

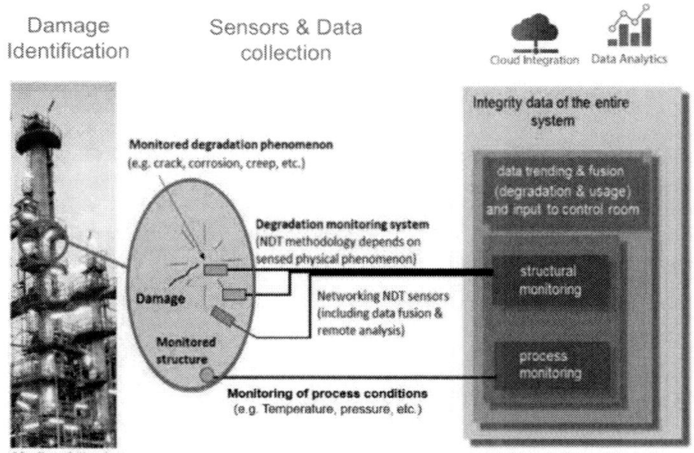


Monitored structure
(e.g. distillation column)

Configured for **auto reporting, auto alarms**

# ASSET HEALTH MONITORING APPROACH – 2/2

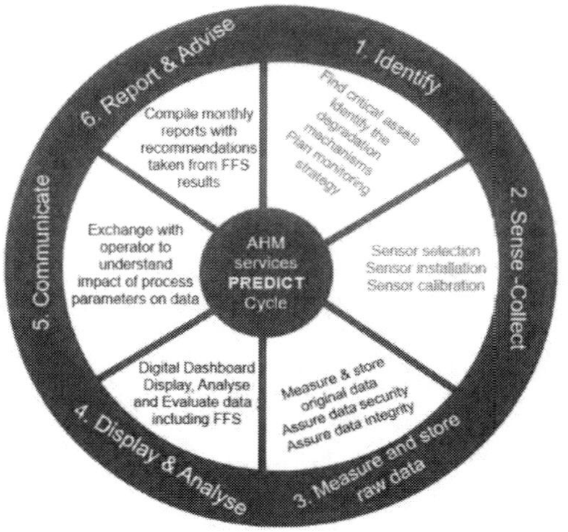




## ASSET HEALTH MONITORING Solution

Sensors for Wall thickness monitoring

- Data transmission: Online or download on PC
- High Temperature Measurement: -30°C to 500°C
- Minimum wall thickness: 3mm (soon 2mm +)
- Resolution: 0.020mm
- Maximum number of probes per PM Box: 8-16
- Storage Capacity: Up to 3000 measurements
- Equipment for ATEX Ex II 3G areas

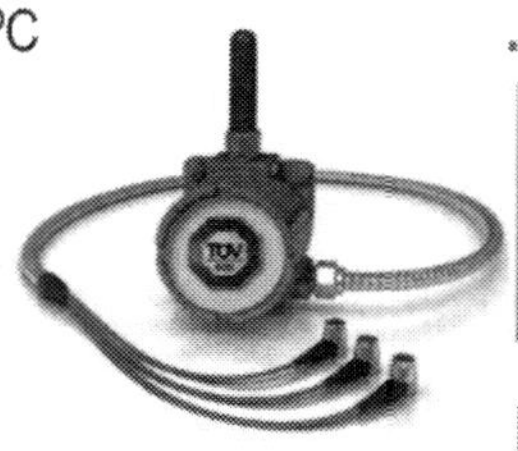

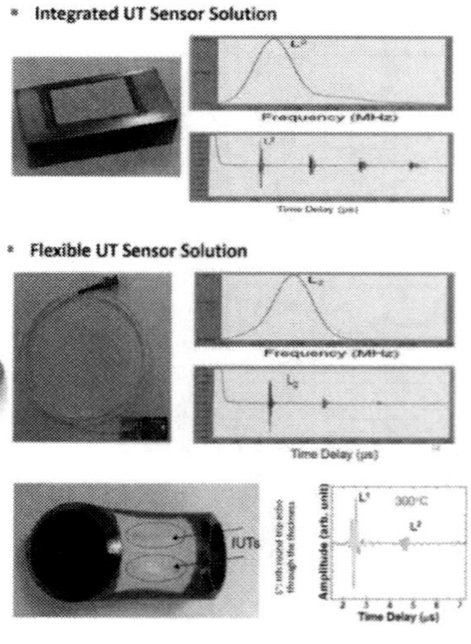

## Value added of ASSET HEALTH MONITORING DATA for Process Safety

### Detecting Impact of processing parameters on Mechanical Integrity

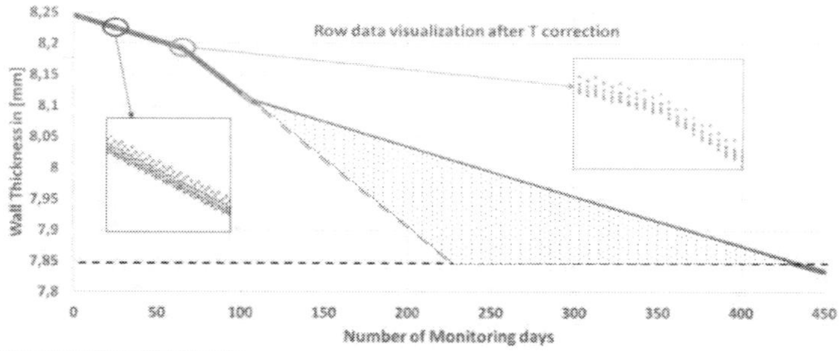


**Find Pattern in BIG DATA to improve Process Safety and achieve higher plant longevity**

**Advantages of AHM system**

- Use BIG DATA for Pattern recognition
- Detect influence of changing process parameters on Mechanical Integrity
- Support optimization of process parameters to achieve higher plant availability & plant longevity
- Inform operators on time before major negative consequences arise
- Support RCM and RBI Approach



## ASSET HEALTH MONITORING – Examples of areas of applications

- Atmospheric Gas-Oil Lines
- Hydro-Fluoric (HF) Alky Units
- Vacuum Fractionators
- Crude Overhead Lines
- Buried pipelines
- CUI and FAC
- Difficult to access locations
- Highly critical assets (probability of failure and consequence of failure)
- Sulfuric Acid Alkylation Units
- Pipe elbows suffering from corrosion and/erosion
- Critical equipment that should not fall out of production
- Assets with high corrosion rates
- Buried pipes and tanks
- Butane line corrosion monitoring
- Sulfur reducing units
- Crude line corrosion monitoring
- Insulated pipes
- Flare Gas Lines

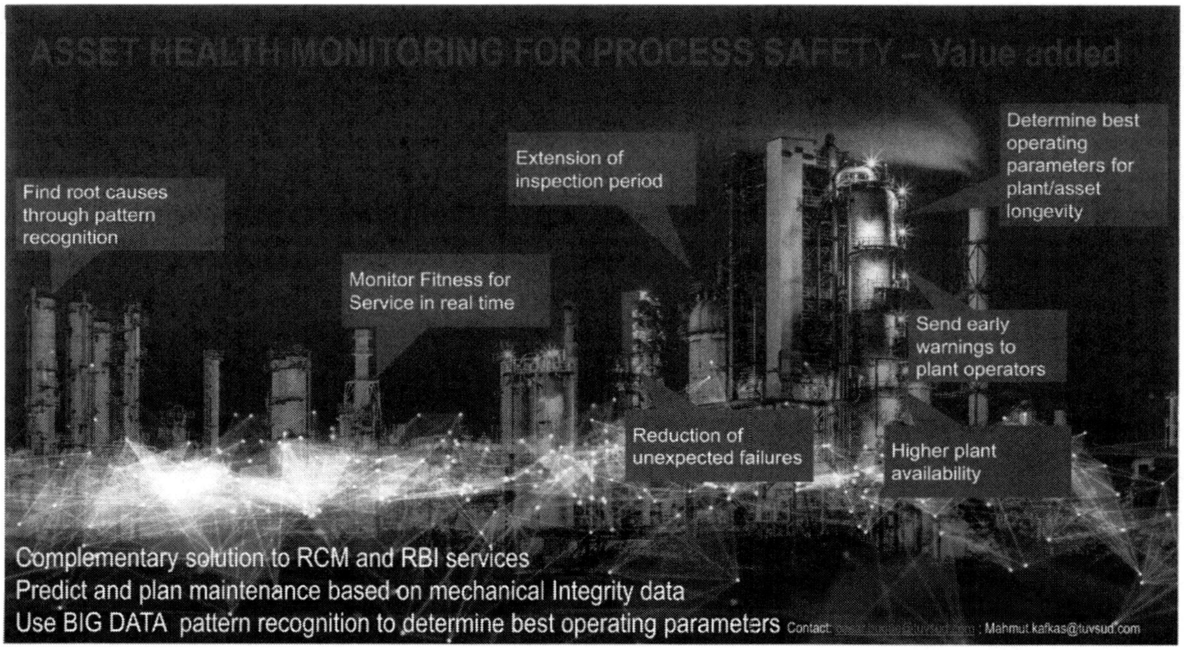


## Contact US


**Dr Cesar Buque**
Global Manager NDT & Mechanical Integrity

TÜV SÜD AG
Division Industry Service

cesar.buque@tuvsud.com

**Mahmut Kafkas**

Manager Innovation & Digitalization

TÜV SÜD AG
Division Industry Service

mahmut.kafkas@tuvsud.com


### Value add of ASSET HEALTH MONITORING Systems

- Quantitative Monitoring of corrosion, erosion, cracking, thermal expansion,
- Monitor Fitness for service real time
- Send early warnings to plant operators according to determined thresholds
- Predict and plan maintenance based on mechanical Integrity data & RBI
- Find root causes through pattern recognition
- Determine best operating parameters for plant/asset longevity
- Achieve higher plant availability & Extension of inspection period

# Process Safety: Understanding how AI can Drive Value from Big Data

Jim Chappell
AVEVA
Head of AI & Advanced Analytics

October 2020

AVEVA

## Industrial AI Overview: Safety

Engineering / Operations / Performance

- Predictive
- Performance
- Prescriptive
- Prognostic
- Perceptive

**Maintain**
Increased longevity and performance of assets while ensuring a safe, reliable environment for the workforce through predictive, prescriptive, ... analytics

**Operate | Optimize**
Improved Optimization & OEE for safe and profitable operations within constraints and regulatory norms

**Plan | Schedule**
Self-learning approach to model planning, scheduling in order to maximize profitability and efficiency

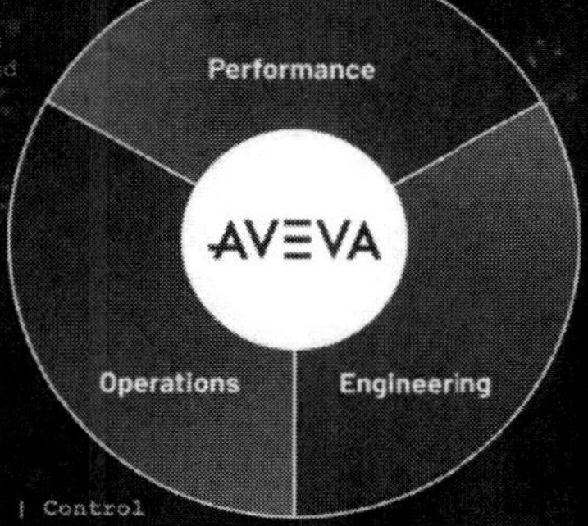


**Engineer**
Analysis and automated design generation enabling lower total cost, time, risk in capital project engineering

**Monitor | Control**
Predictive and autonomous operational control to ensure safety, performance

AVEVA

## Enhanced Analytics for Reduced Downtime & Safety

**Failure Patterns**

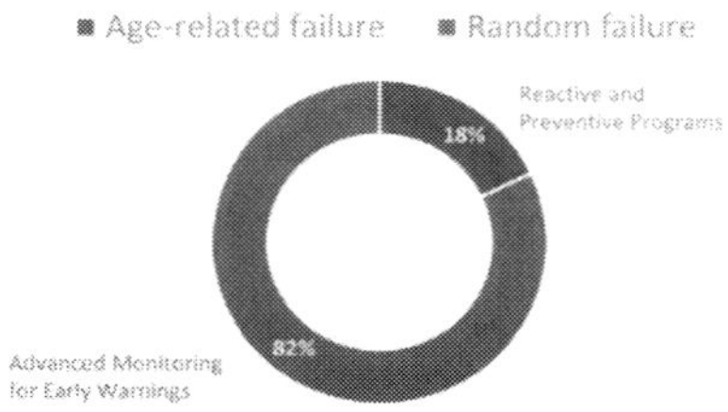


ARC estimates that unscheduled downtime costs the process industries somewhere in the area of a trillion dollars a year

$1,000,000,000,000

**It's a Journey**

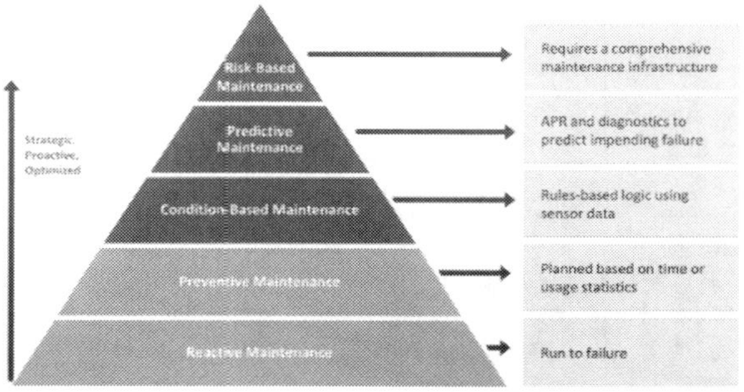


ARC STUDIES SHOW ONLY 18% OF ASSET FAILURE IS AGE-RELATED. BASED ON THESE DATA, PREVENTIVE MAINTENANCE PROVIDES A BENEFIT FOR JUST 18 PERCENT OF ASSETS, AND MONITORING FOR PREDICTIVE MAINTENANCE IS A RECOMMENDED OPTION FOR THE REST. WWW.ARCWEB.COM/LISTS/POSTS/POST.ASPX?ID=[illegible]



AVEVA

## Industrial Safety

82% of failures are random and usually not identified ahead of time

⬇

Equipment failures lead to unstable operations

⬇

Unstable operations lead to operational risk and process safety issues



AVEVA

## Improved Safety with AI

**Mitigate / eliminate process safety issues by:**

1. Identifying potential problems well before they happen
2. Assessing impact and risk from all angles

*Safety issues often stem from a lack of planning and analysis*



AVEVA

## Integrated AI for Safety

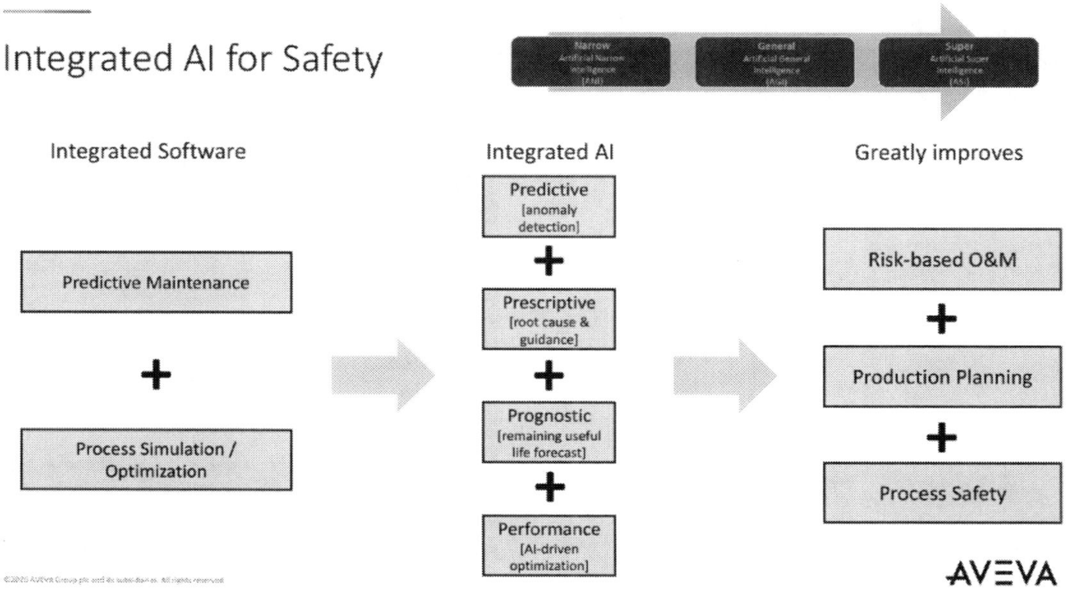




AVEVA

# AI-driven Asset Performance Management

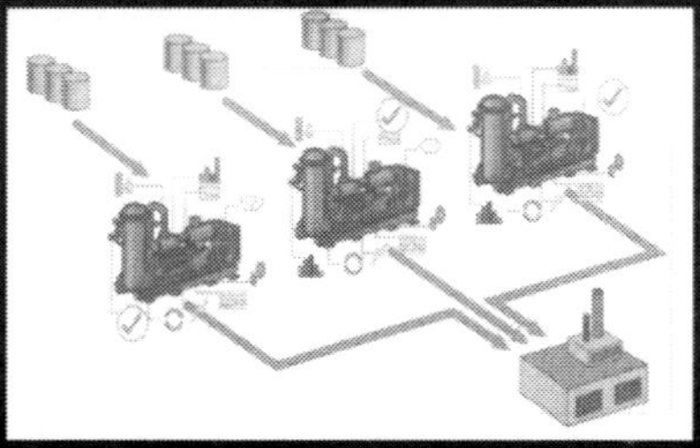

**Historical**

What Happened

- Assessment and exploration of historical operational data
- Trends, KPIs, Dashboards to present abstracted views

*Historical Domain*

**Real-Time**

What's Happening

- Processing of real-time operational data
- Rule based inference for causal analysis
- First principles

*Real-Time Domain*

**Predictive**

What If

- Comprehensive model based assessment of operational data ranges to determine potential outcomes.
- Deterministic or non-deterministic models
- Early detection of issues

*Machine Learning Domain*

**Prescriptive**

What to do

- Systems that synthesize, predict and provide scenario-based guidance
- Fault diagnostics and knowledge capture
- Sensor contribution & root-cause analysis
- Extensive library of prescriptive actions

*Maintenance Action Domain*

**Prognostics**

How bad will it get

- Forecast future state of assets and sensor values
- Determine if you can make it to the next planned maintenance outage
- Provide key input for risk assessment

*Artificial Intelligence Domain*

AVEVA

# Add AI-infused Simulation/Optimization

Game-changing Integrated AI and Simulation

Asset Performance

Operate and Optimize

**High Dynamic Range Models (earlier detection of issues)**

**Risk-based AI Analysis**

- Impact
- Optimization

**Single Enterprise Notification System**

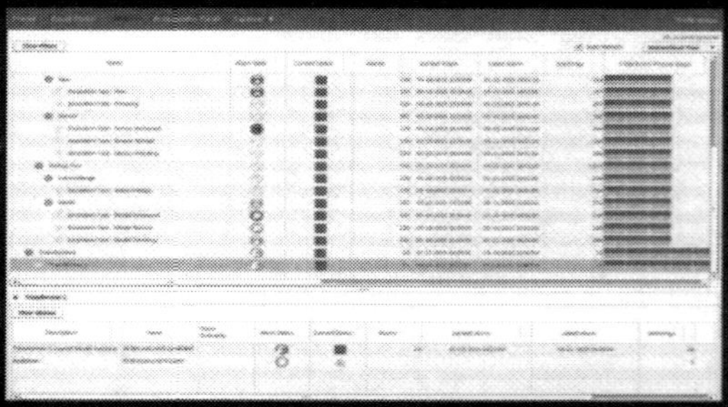

AVEVA

## Real-world: Boiler Safety

**Observation:**

- An operator changed the draught controls to manual and forgot to change it back to auto while safety interlocks were not properly set. Then the load was increased causing an increase in fuel supplied with a correspondingly improper O2 level.

**Result:**

- According to plant personnel, if AI-driven analytics hadn't caught the issue, there was a significant chance for either a major unplanned outage or an explosion.

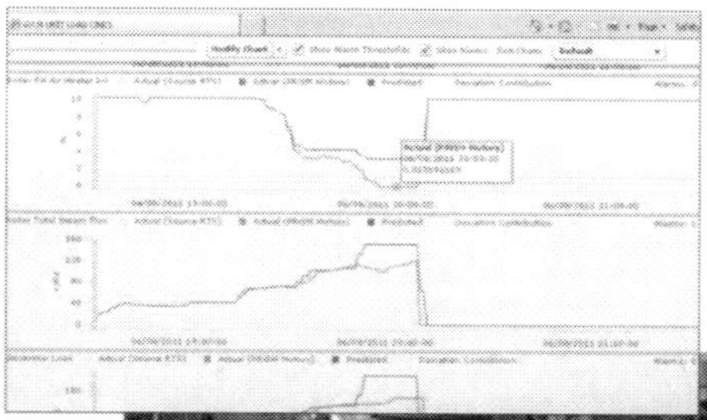

## Real-world: Transformer Safety

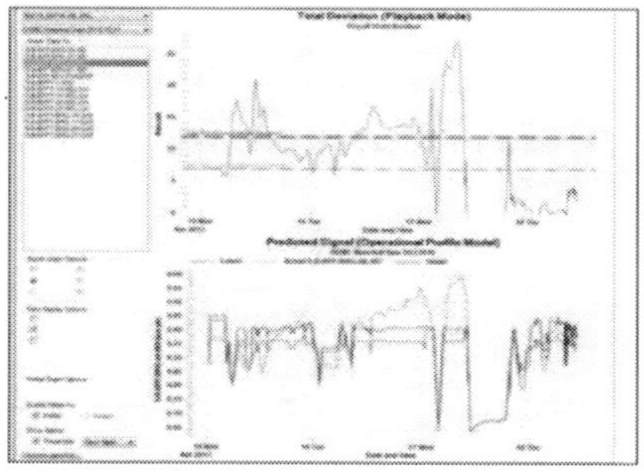

**Observation:**

- A major power company found a high-voltage transformer issue during Hurricane Matthew in October 2016. One of the predictive transformer health models alarmed, indicating problems. The issue was investigated, finding a transformer that was fully charged with no load.

**Result:**

- An explosion could have occurred in this situation if loads were brought on too quickly. The AI-driven system successfully prevented a major catastrophe with potential loss of human life.

## Real-world: Compressor Safety

**Observation:**

- Higher than expected vibrations on the second stage of a compressor in relation to the current level of operation.

**Result:**

- The site found a cracked impeller.
- Based on the speed of the crack propagation and simulated impact analysis, the AI system detected this issue approximately 3 months before operators would have noticed it, preventing significant safety issues.

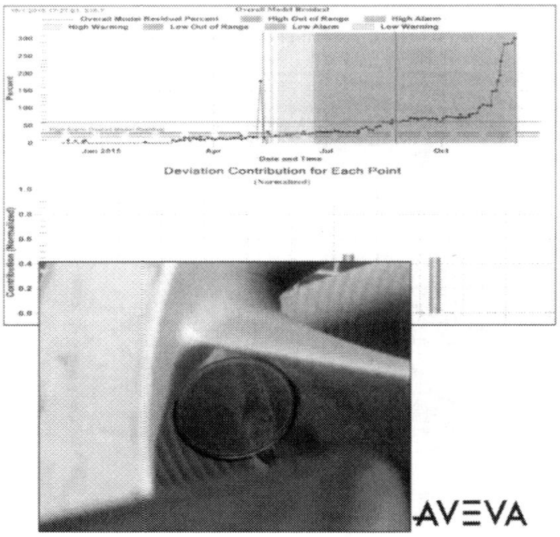




AVEVA

## Real-world: Environmental Safety

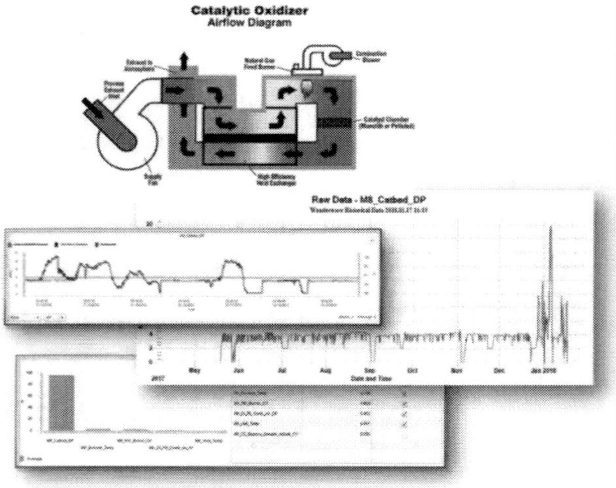


**Observation:**

- An oxidizer is responsible for reducing environmentally harmful emissions by burning the exhaust gases at a high temperature (oxidizing) and then running the hot exhaust over a catalyst to help induce a chemical reaction.

**Result:**

- The predictive system identified a higher than expected differential pressure across the catalyst bed, indicating a clog which can disrupt the oxidation process. However, this issue only occurred once in awhile.
- Maintenance personnel investigated and found that the lines that bleed moisture to the outside were periodically freezing due to weather conditions. They unthawed the lines, allowing the moisture to dissipate which, in turn, reduced the pressure and eliminated the environmental issue.



AVEVA

Thank You!!
AVEVA

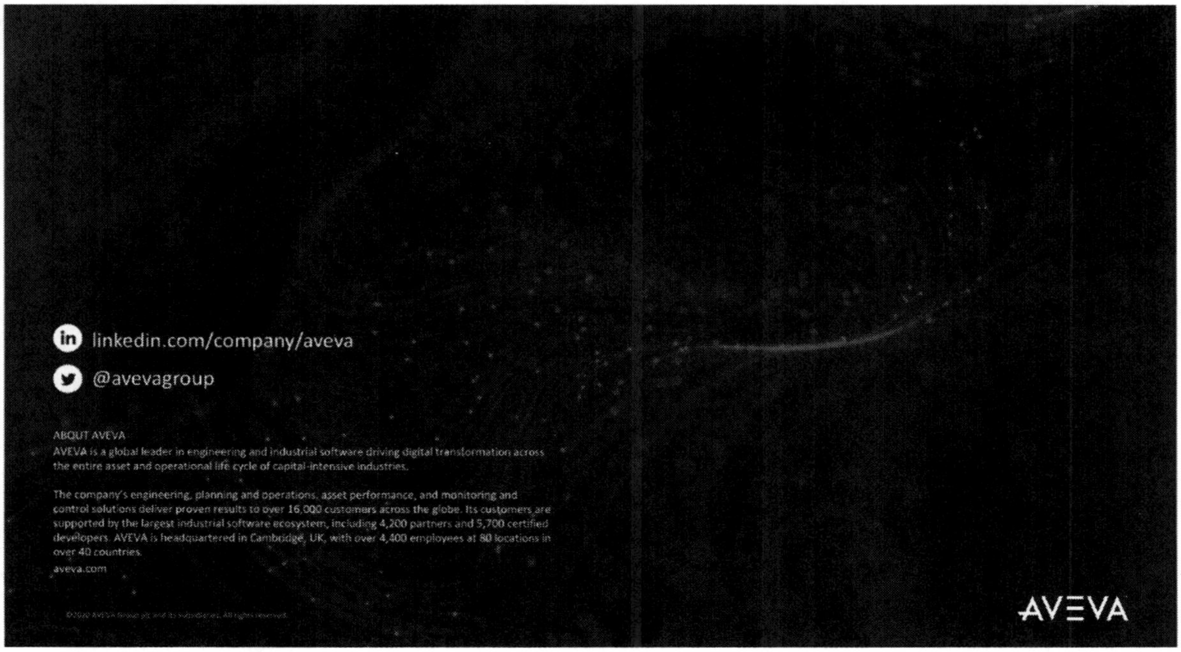
linkedin.com/company/aveva
@avevagroup
ABOUT AVEVA
AVEVA is a global leader in engineering and industrial software driving digital transformation across the entire asset and operational life cycle of capital-intensive industries.
The company's engineering, planning and operations, asset performance, and monitoring and control solutions deliver proven results to over 16,000 customers across the globe. Its customers are supported by the largest industrial software ecosystem, including 4,200 partners and 5,700 certified developers. AVEVA is headquartered in Cambridge, UK, with over 4,400 employees at 80 locations in over 40 countries.
aveva.com
©2020 AVEVA Group plc and its subsidiaries. All rights reserved.
AVEVA

2020 EUROPEAN CONFERENCE ON PROCESS SAFETY & BIG DATA

CCPS EPSC

TAUROB ROBOTIC INSPECTION

# Ground Robots can Lead to a Step Change in Safety

Duco Drenth - Manager Robotics, Dietsmann

Matthias Biegl - Director, Taurob

SESSION 4: Robotic & Emerging Technologies

14 October 2020, 17:40 - 18:30

TAUROB ROBOTIC INSPECTION

# Content

- Intro: Dietsmann & Taurob
- Safety by Design
- Which (Industrial) Robots
- Use Cases

----

- Backup Sheets

Duco Drenth

Matthias Biegl

Dietsmann Maintaining Energy

www.Dietsmann.com

www.Taurob.com

TAUROB ROBOTIC INSPECTION

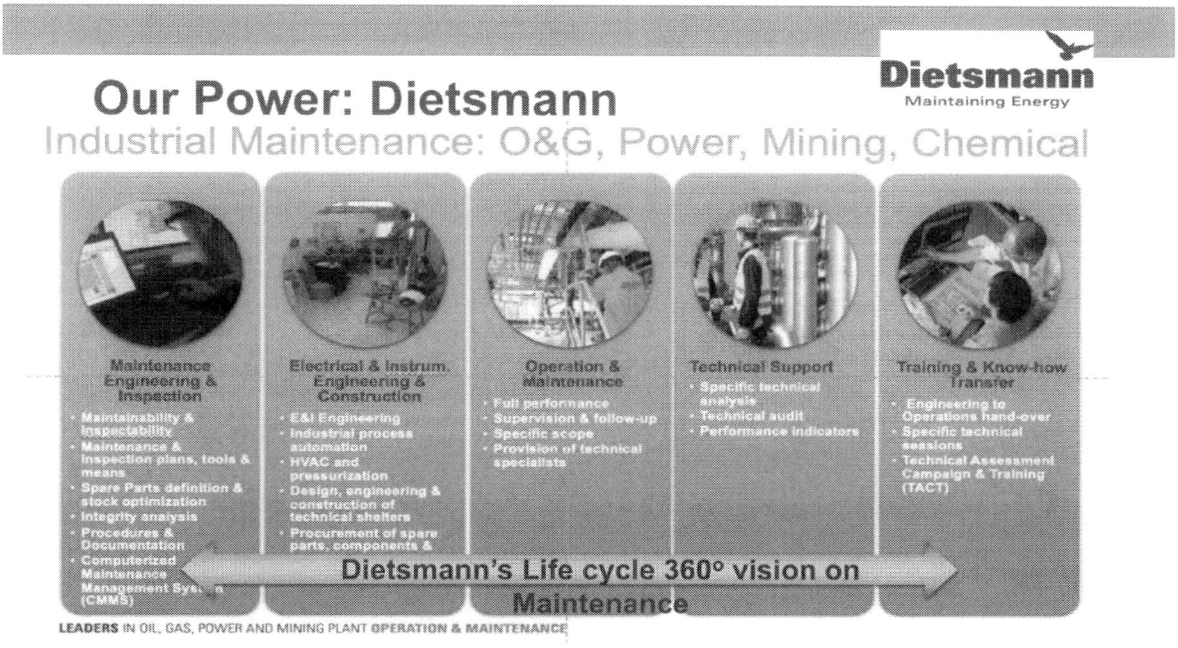
Dietsmann
Maintaining Energy
Our Power: Dietsmann
Industrial Maintenance: O&G, Power, Mining, Chemical
Maintenance Engineering & Inspection
Maintainability & Inspectability
Maintenance & inspection plans, tools & means
Spare Parts definition & stock optimization
Integrity analysis
Procedures & Documentation
Computerized Maintenance Management System (CMMS)
Electrical & Instrum. Engineering & Construction
E&I Engineering
Industrial process automation
HVAC and pressurization
Design, engineering & construction of technical shelters
Procurement of spare parts, components &
Operation & Maintenance
Full performance
Supervision & follow-up
Specific scope
Provision of technical specialists
Technical Support
Specific technical analysis
Technical audit
Performance indicators
Training & Know-how Transfer
Engineering to Operations hand-over
Specific technical sessions
Technical Assessment Campaign & Training (TACT)
Dietsmann's Life cycle 360° vision on Maintenance
LEADERS IN OIL, GAS, POWER AND MINING PLANT OPERATION & MAINTENANCE

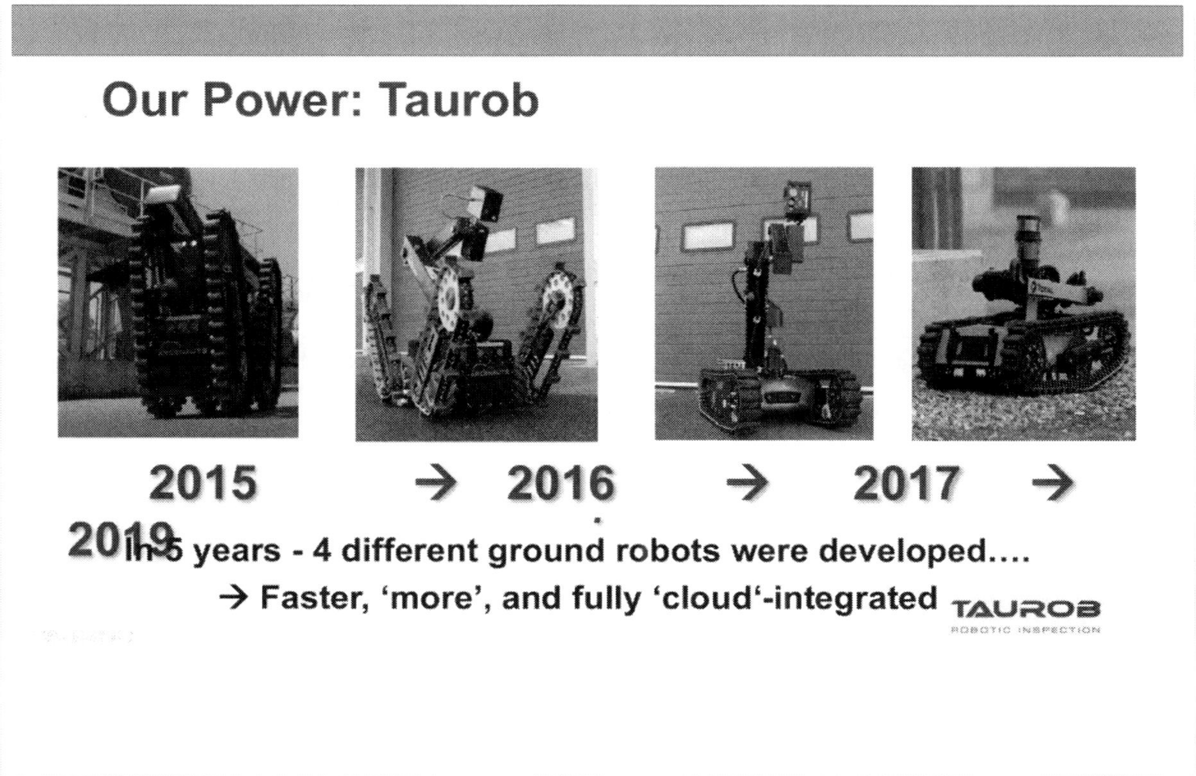
Our Power: Taurob
2015 → 2016 → 2017 → 2019
In 5 years - 4 different ground robots were developed....
→ Faster, 'more', and fully 'cloud'-integrated
TAUROB
ROBOTIC INSPECTION

# Joint Power

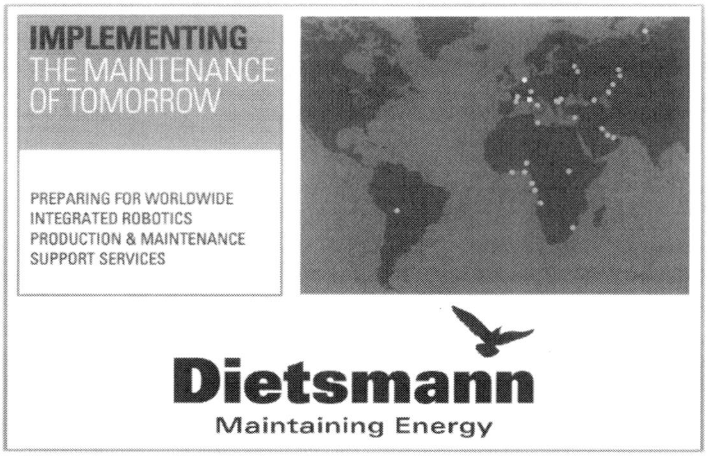


**Dietsmann**

- Operation & Maintenance
- 40+ years experience
- 20+ affiliates

**Taurob**

- Robots & Innovation
- 10+ years experience
- 1 affiliate

TAUROB
ROBOTIC INSPECTION

# Safety By Design

We only design, plan, or perform our work in such a manner as to prevent foreseeable accidents from occurring, by eliminating or minimising the risks of (any) harm.

1. *Elimination (physically remove the hazard).*
2. *Substitution (replace the hazard).*
3. *Engineering Controls (isolate people from the hazard).*
4. *Administrative Controls (change the way people work).*
5. *PPE (protect worker with Personal Protective Equipment).*

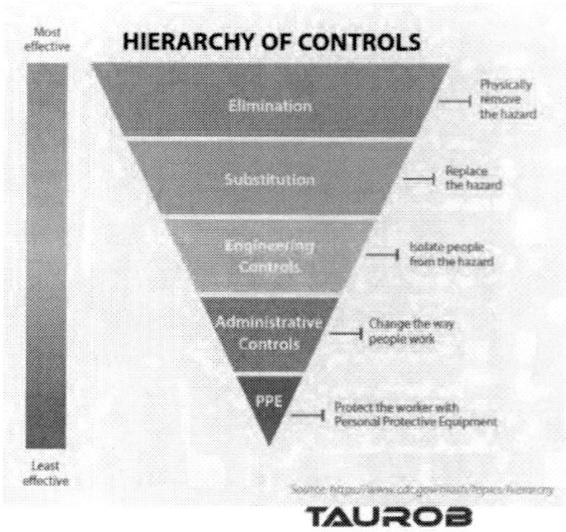


TAUROB
ROBOTIC INSPECTION

## Safety By Design

We only design, plan, or perform our work in such a manner as to prevent foreseeable accidents from occurring, by eliminating or minimising the risks of (any) harm.

1. *Elimination (physically remove the hazard).*
2. *Substitution (replace the hazard).*
3. *Engineering Controls (isolate people from the hazard).*
4. *Administrative Controls (change the way people work).*
5. *PPE (protect worker with Personal Protective Equipment).*

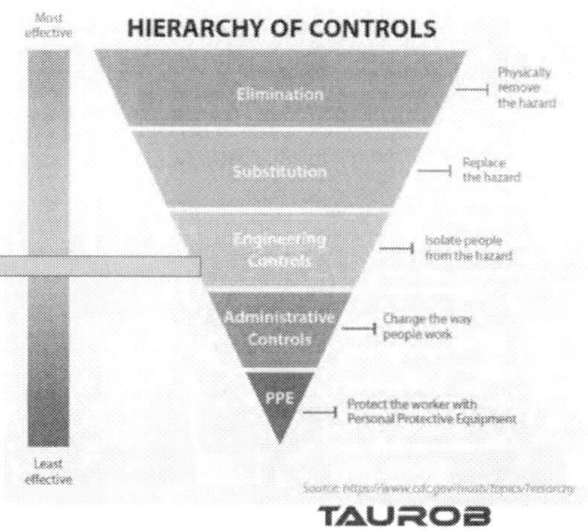


TAUROB
ROBOTIC INSPECTION

## Safety By Design

**Some work performed by humans <u>can</u> be replaced by robots:**

- Avoid human entry of confined spaces (e.g. tanks).
- Hazardous tasks: working at height, fire fighting, etc.
- Access to 'inaccessible' places (e.g. no scaffolding).
- Less repetitive and monotonous tasks (for humans).
- Replace human operator at remote and offshore sites.
- Less transport of people.
- Reduce waste (e.g. cleaning, emptying tanks), avoid leakage
- (spill control) and limit catastrophic events (e.g. clean-up).

TAUROB
ROBOTIC INSPECTION

- If autonomous: more frequent monitoring & inspection, with

# We Share Your Commitment for Safety

Improve safety, effectiveness and working conditions for our clients and their employees

Keep attractive, safe, working conditions for staff, either on-site or off-site.

Robots that operate and co-work safely with humans

TAUROB
ROBOTIC INSPECTION

# Benefits to Your Operations

1 **Improve consistency / coherence of data to support & (continuously) improve asset performance**

- Data gathering on equipment status - efficiency & effectiveness
- Assist workforce in day-to-day activities - efficiency
- Reduce Emissions (e.g. unplanned: leaks) - HSE, efficiency
- Logistics - HSE, efficiency

2 **Reduce Risks (incl. incidents)**

- Enhance safety (environment & health) - HSE

3 **Implement 'Maintenance of the Future': *'Operator 4.0'***

- Predictive Maintenance - reliability
- Preparedness for unexpected events - reliability

Allowing to reduce OPEX, and CAPEX, while assuring

TAUROB
ROBOTIC INSPECTION

# What (Else) Do We Need?

→ ***Ground Robots are THE Data Collectors…..***

- *Onboard **Equipment/ Sensors**:*

→ *(HD, Infrared) Camera (vision, leakage).*

→ *(Ultra) Sound recorder (frequency, volume,..).*

→ *LIDAR (3D-scan), GPS (location).*

→ *Heat/ Temperature.*

→ *Gas Detection ($CH_4$, $CO_2$, $H_2S$, …).*

→ *Chemicals (e.g. pH).*

→ *Other, e.g. corrosion inspection, radiation,….*

**24/7…**

→ **Plus: light, data-transmission/ base station,, …real-time**

**Additional** Ears, Eyes, Nose Limbs, Brains

ROBOTIC INSPECTION

# A Wide Array of Use Cases

| | |
|---|---|
| 1 | Scouting, Situational awareness |
| 2 | Routine inspection |
| 3 | Autonomous missions |
| 4 | Standard operations |
| 5 | Intervention operations |
| 6 | Emergency response |
| 7 | Standard maintenance/ manipulation |
| 8 | Complex analyses (sound / picture /…) |
| 9 | Data capture and streaming |
| 10 | Capture asset integrity over time |

→ Predictive Maintenance analyses……

TAUROB
ROBOTIC INSPECTION

Signalling abnormalities

$360^0$ images and video

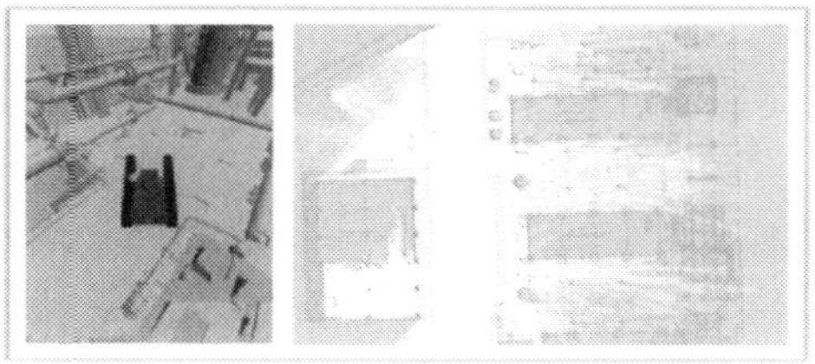

Scanning of location, modelling

Reading gauges, Status checks

Vibration monitoring

Autonomous Missions, Unmanned Platforms

# Robot Capabiliti es

TAUROB
ROBOTIC INSPECTION

# User Interface: Simple, Portable, Powerful

- Runs on ATEX Tablets (Android) and Big Screen PCs
- Allows Mission Planning and Supervision
- Driving via Touch Gestures

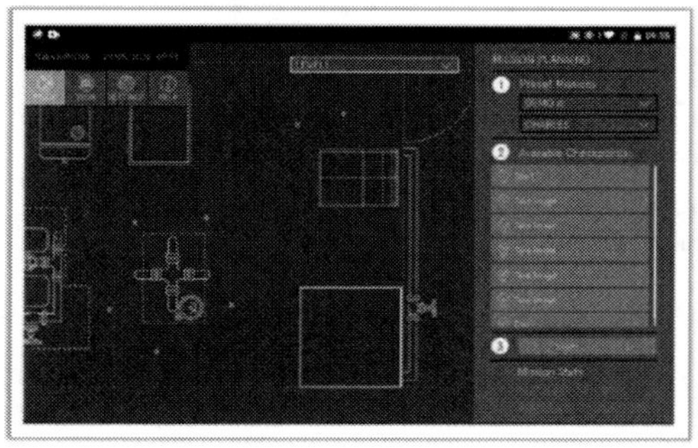

TAUROB
ROBOTIC INSPECTION

ROBOTIC INSPECTION

# Safety By Design

We only design, plan, or perform our work in such a manner as to prevent foreseeable accidents from occurring, by eliminating or minimising the risks of (any) harm.

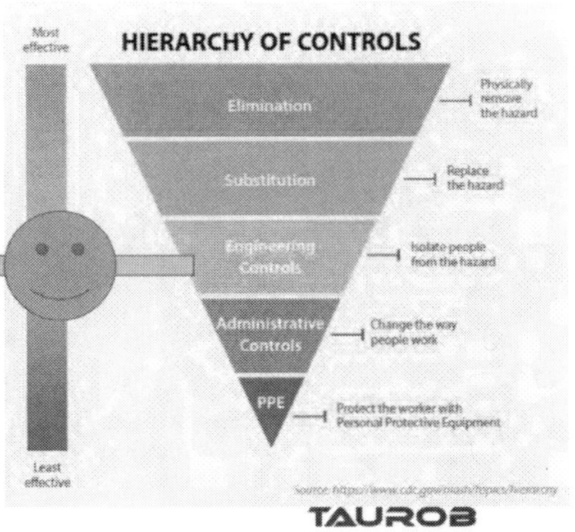


TAUROB
ROBOTIC INSPECTION

# In All Kinds of Conditions

*'No Maintenance, No Products'*

TAUROB
ROBOTIC INSPECTION

# TAUROB

ROBOTIC INSPECTION

# *Thank you.*

All Slides are Taurob Proprietary and Confidential Information.

Photo Credits: taurob GmbH
Total (Laurent Pascal)
Total (Richard Davies)
Total (Ben Mullay)

Dietsmann
Maintaining Energy

TAUROB
ROBOTIC INSPECTION

# BACK-UP SHEETS

TAUROB
ROBOTIC INSPECTION

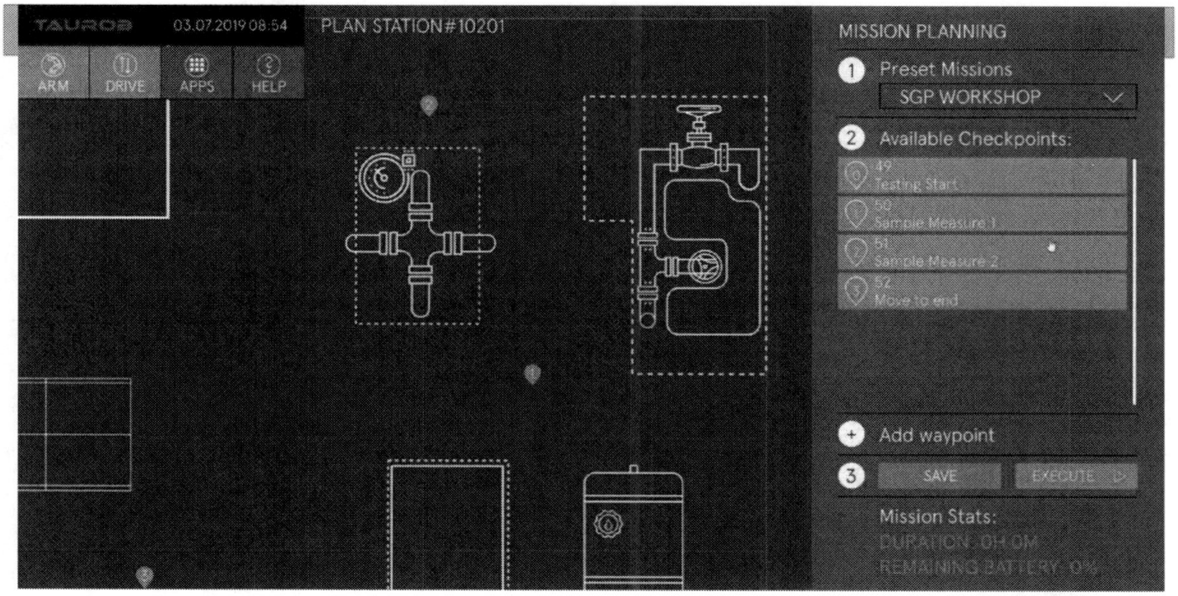


TAUROB
ROBOTIC INSPECTION

ROBOTIC INSPECTION

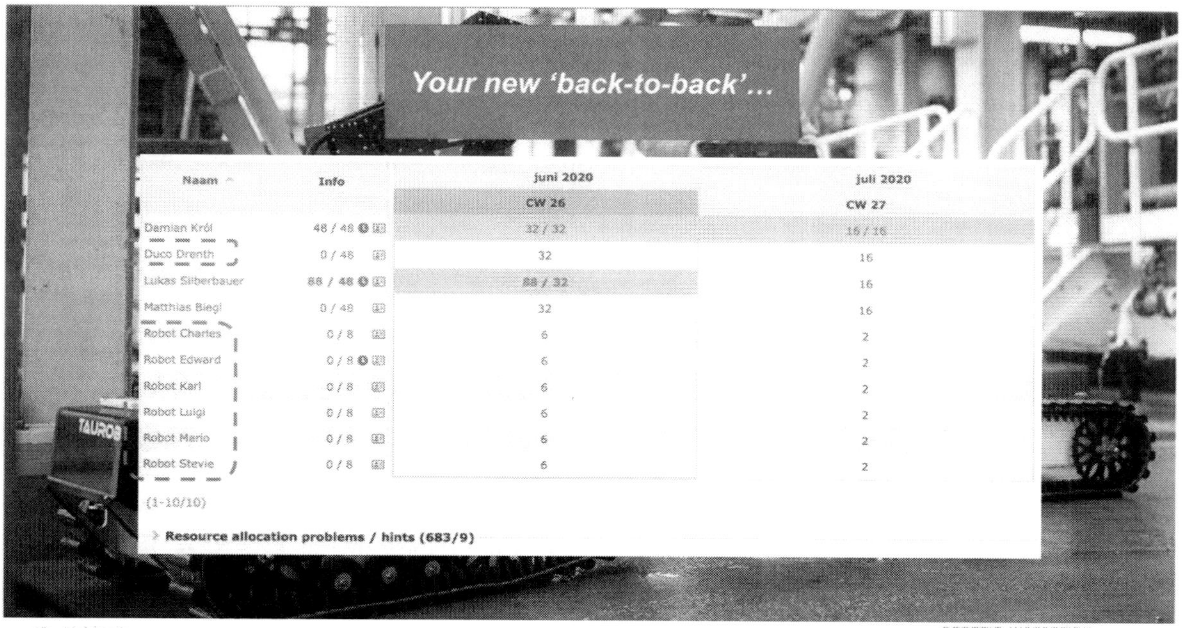


| Naam | Info | juni 2020 CW 26 | juli 2020 CW 27 |
|---|---|---|---|
| Damian Król | 48 / 48 | 32 / 32 | 16 / 16 |
| Duco Drenth | 0 / 48 | 32 | 16 |
| Lukas Silberbauer | 88 / 48 | 88 / 32 | 16 |
| Matthias Biegl | 0 / 48 | 32 | 16 |
| Robot Charles | 0 / 8 | 6 | 2 |
| Robot Edward | 0 / 8 | 6 | 2 |
| Robot Karl | 0 / 8 | 6 | 2 |
| Robot Luigi | 0 / 8 | 6 | 2 |
| Robot Mario | 0 / 8 | 6 | 2 |
| Robot Stevie | 0 / 8 | 6 | 2 |

(1-10/10)

› **Resource allocation problems / hints (683/9)**

ROBOTIC INSPECTION

## Remote Working is <u>not</u> the Future....

TAUROB
ROBOTIC INSPECTION

# The Total Picture….

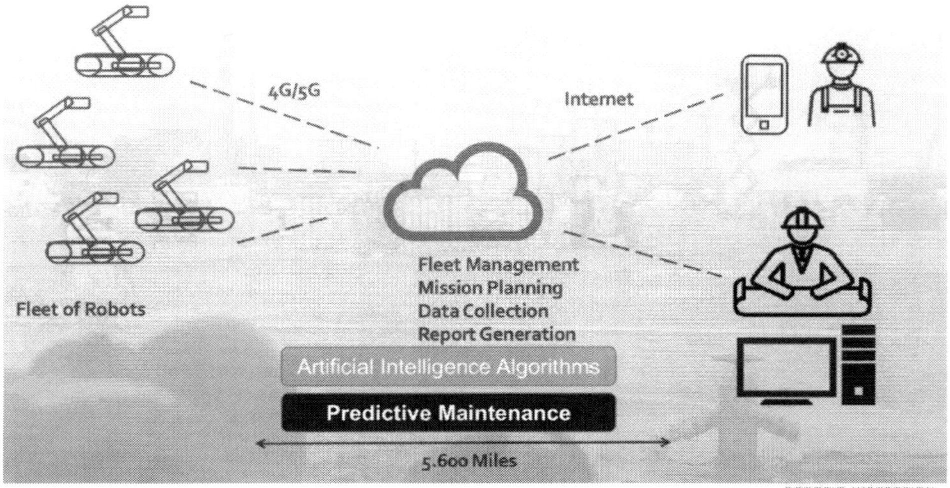


ROBOTIC INSPECTION

ROBOTIC INSPECTION

TAUROB

Dietsmann
Maintaining Energy

ROBOTIC INSPECTION

# Proof of Concept

TAUROB

ROBOTIC INSPECTION

# Please note

IBM's statements regarding its plans, directions, and intent are subject to change or withdrawal without notice at IBM's sole discretion.

Information regarding potential future products is intended to outline our general product direction and it should not be relied on in making a purchasing decision.

The information mentioned regarding potential future products is not a commitment, promise, or legal obligation to deliver any material, code or functionality. Information about potential future products may not be incorporated into any contract.

The development, release, and timing of any future features or functionality described for our products remains at our sole discretion.



2

# Key Takeaways

- What is changing for maintenance technicians and companies?
- Maximo Solutions and Capabilities
- Case Studies
- How can you get started?
- Q & A




# Field Reinvention is driven by technology and demographics

## Equipment is Becoming Smarter, More Sophisticated and More Connected

Operational and Maintenance Data exists across multiple systems

More Digitized Information About Equipment is Available through IoT, Digital Twin, etc.

Cloud Computing & Big Data Allow Massive Amounts of Information to be Aggregated

## Changing workforce requires next-generational technology

Aging Workforce and Loss of Tribal Knowledge

Digital native audience expects consumer-grade capabilities

AI Technology is accessible to the technicians to help increase productivity

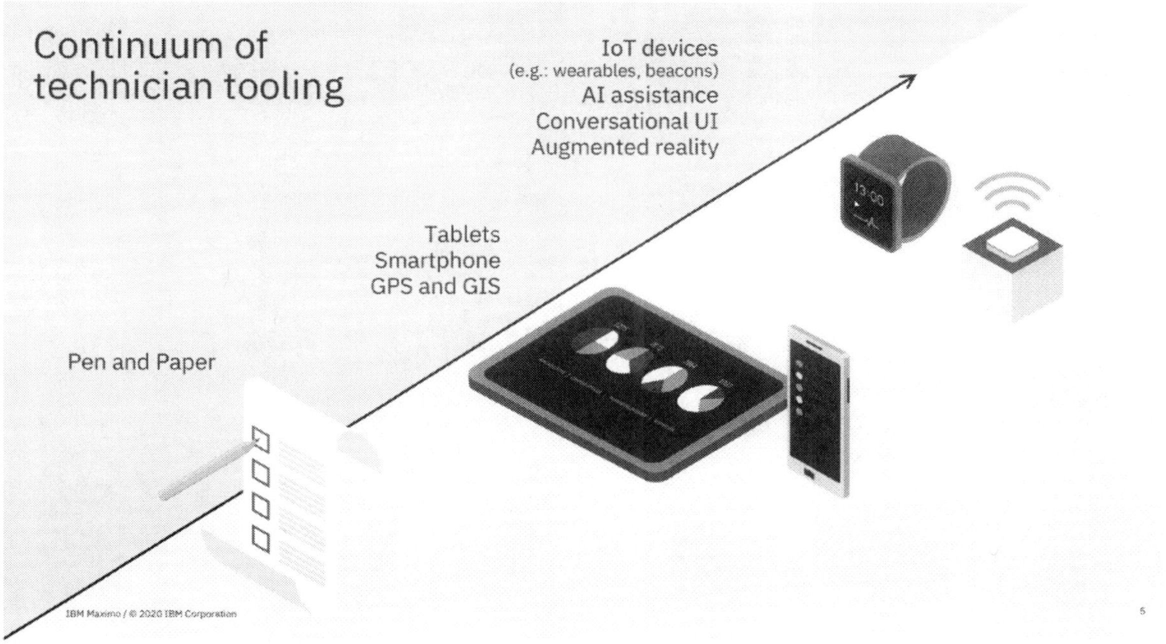


## Maintenance Technician
Leaders vs Laggards

**Laggards**

- Aging Workforce Problem
- Pen and Paper world
- No standardized repair processes (e.g.: shotgun approach)
- Unnecessary repair and parts replacement
- Data is in disparate silos or tribal knowledge
- High overall maintenance costs

**No significant change in the way their field service teams operate**

**Leaders**

- Tools for digital native workforce
- Reliability centered maintenance
- Leverage expertise and share knowledge across organization
- Reduced mean time to action and improved first time fix rate

**Technicians are more productive, safer and in compliance**

**Organizations can reduce maintenance and operational costs**


6

The need is urgent
1.5X
25%
20%
Turnover
Retiring
Hiring Cost
3rd party Techs
Fail 1st Fix
Reduce Operating Costs
Attract and Retain Talent
Improve Customer Engagement
Increase Worker Safety

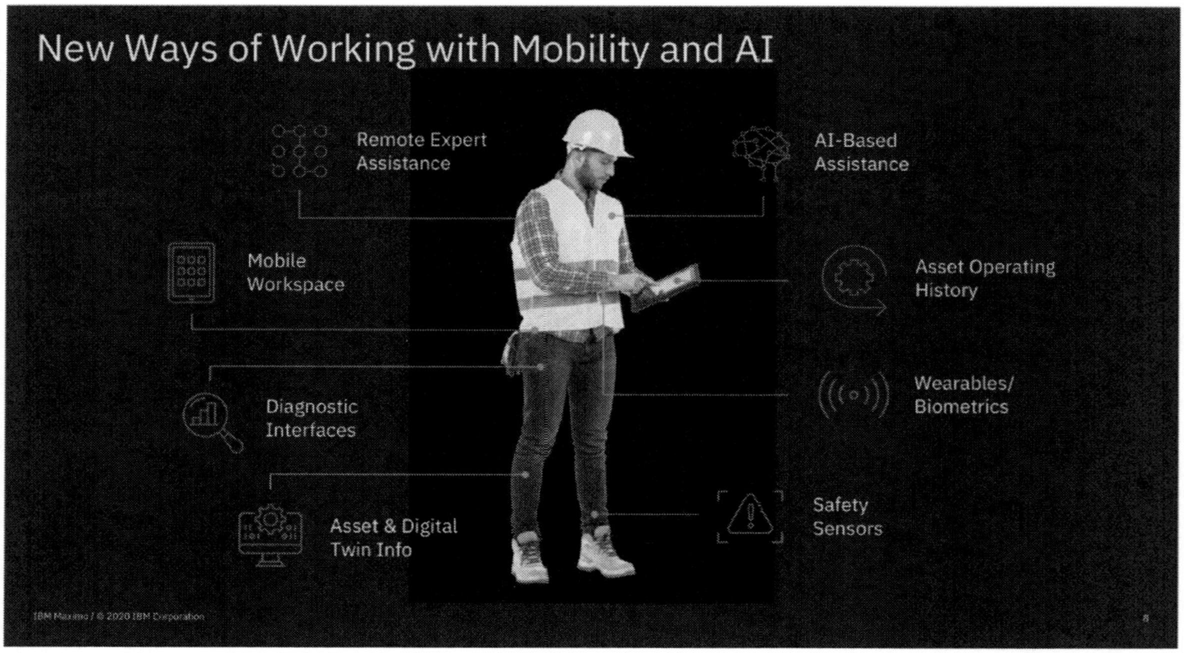
New Ways of Working with Mobility and AI
Remote Expert Assistance
AI-Based Assistance
Mobile Workspace
Asset Operating History
Wearables/ Biometrics
Diagnostic Interfaces
Safety Sensors
Asset & Digital Twin Info

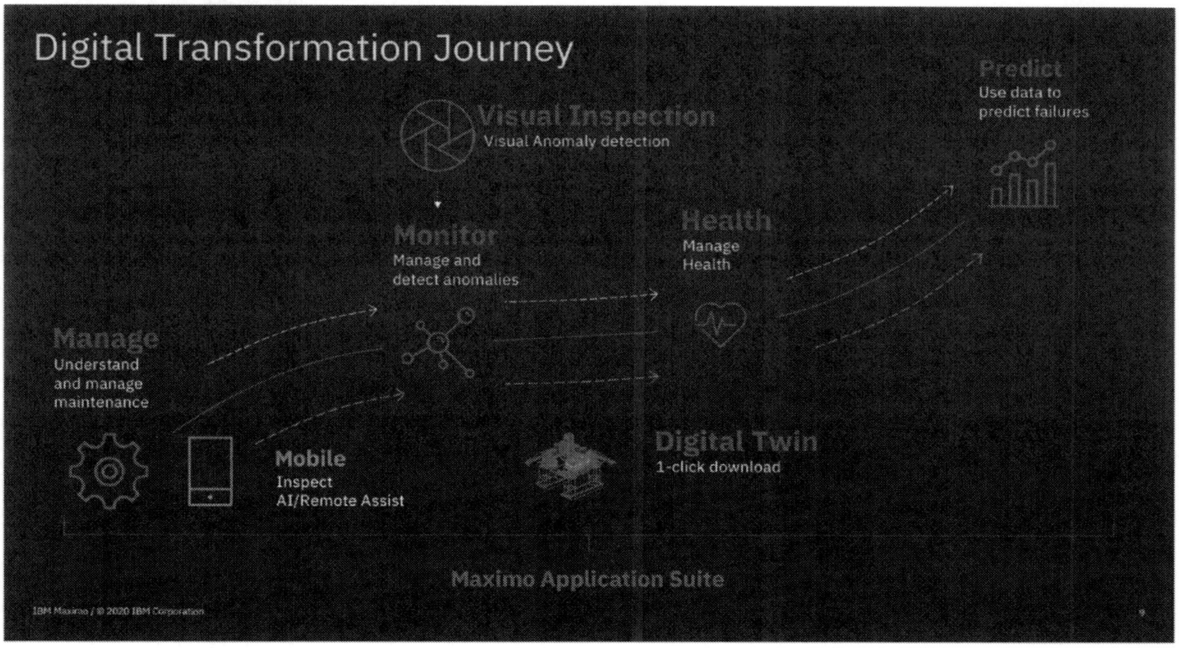
Digital Transformation Journey
Predict
Use data to predict failures
Visual Inspection
Visual Anomaly detection
Monitor
Manage and detect anomalies
Health
Manage Health
Manage
Understand and manage maintenance
Mobile
Inspect
AI/Remote Assist
Digital Twin
1-click download
Maximo Application Suite
IBM Maximo / © 2020 IBM Corporation
9

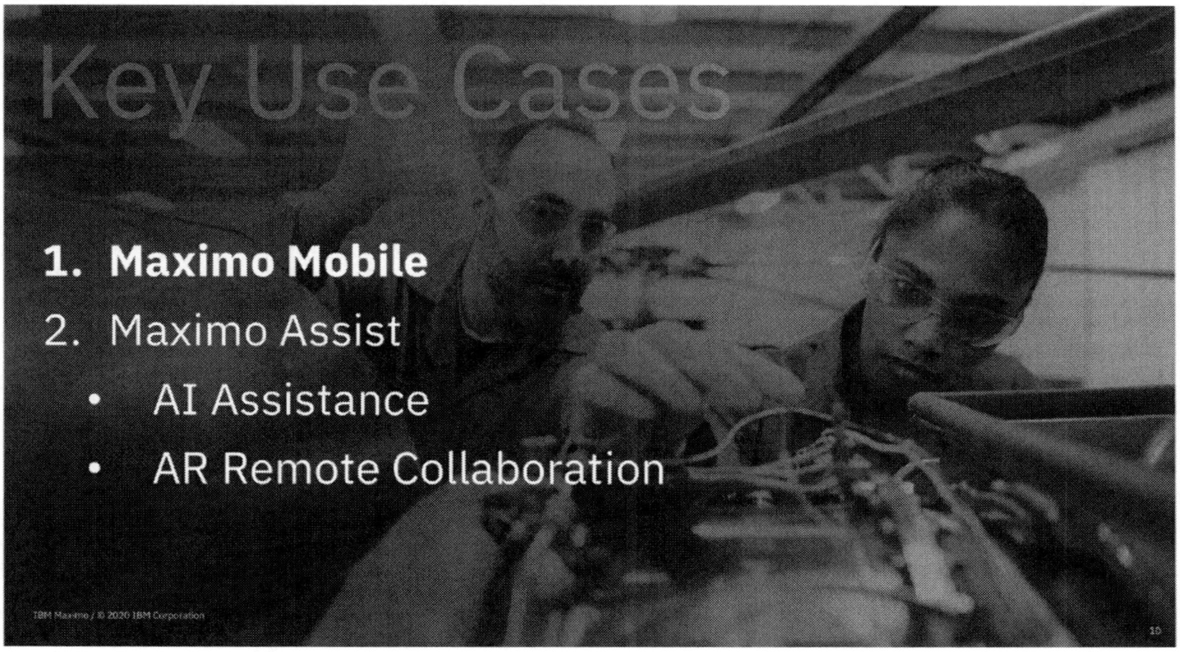
Key Use Cases
1. Maximo Mobile
2. Maximo Assist
• AI Assistance
• AR Remote Collaboration
IBM Maximo / © 2020 IBM Corporation
10

# Maximo Mobile

Next Generation User Experience

- 1 application, 1 role-based interface
- Intelligent workflows = blue button
- Intelligent forms = configurability
- Connected and disconnected
- Trustworthiness: data quality is higher, and real-time updates

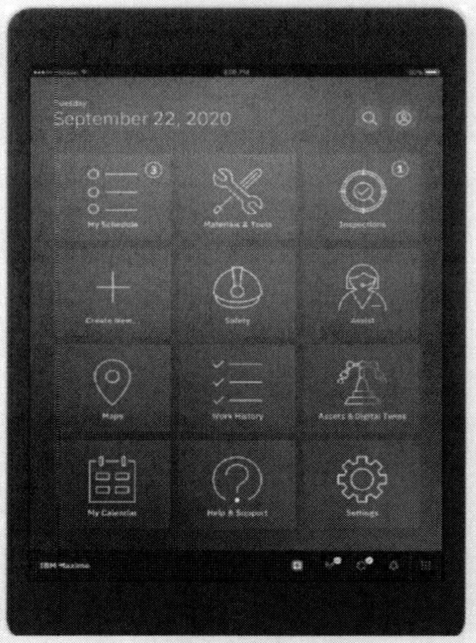




# Maximo Mobile

**Next Generation User Experience** that's simple to use

Just click on the blue button "guided" flow

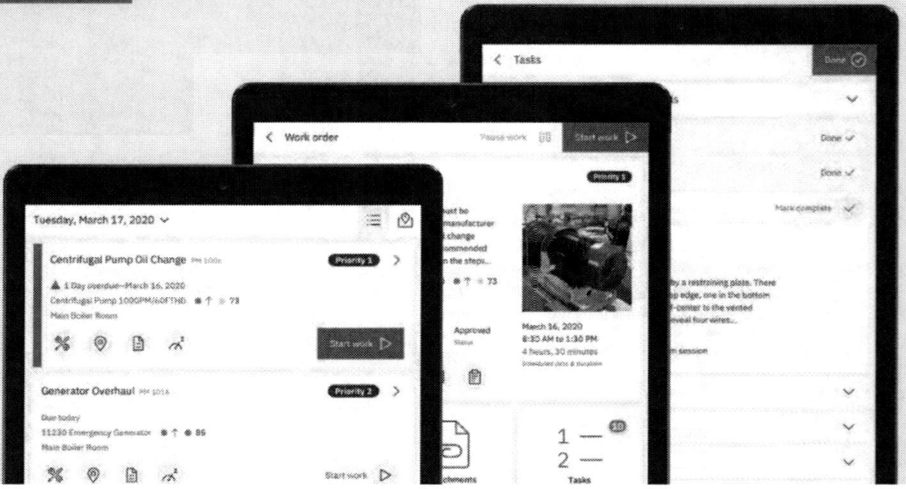

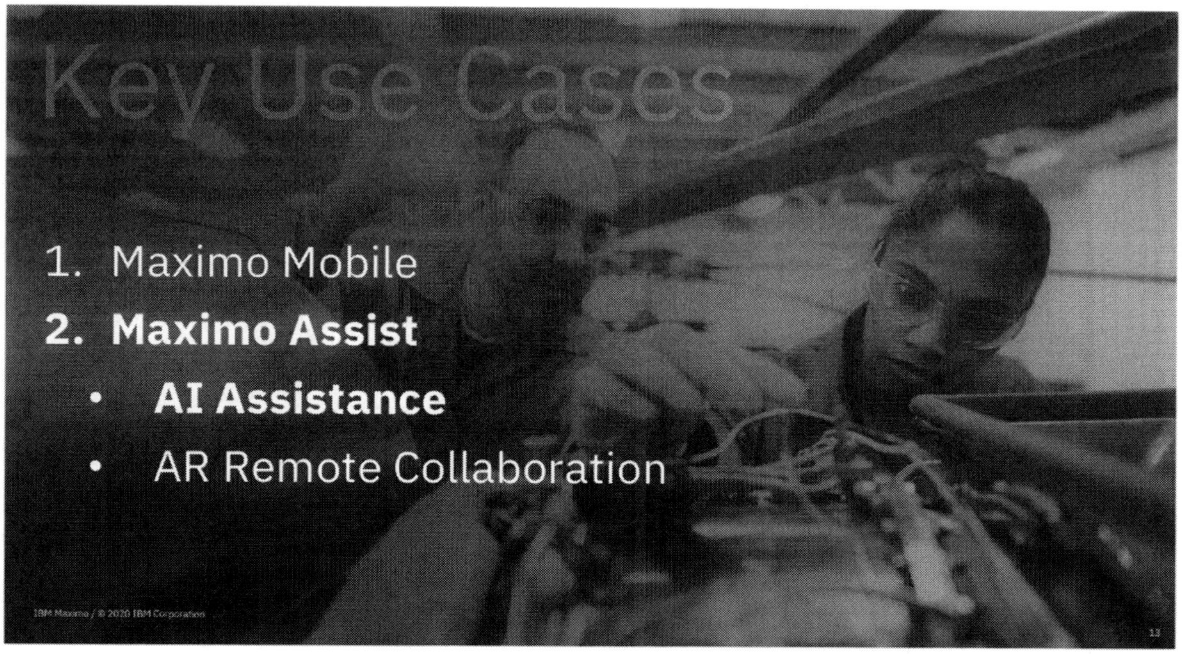
Key Use Cases
1. Maximo Mobile
2. Maximo Assist
• AI Assistance
• AR Remote Collaboration
IBM Maximo / © 2020 IBM Corporation
13

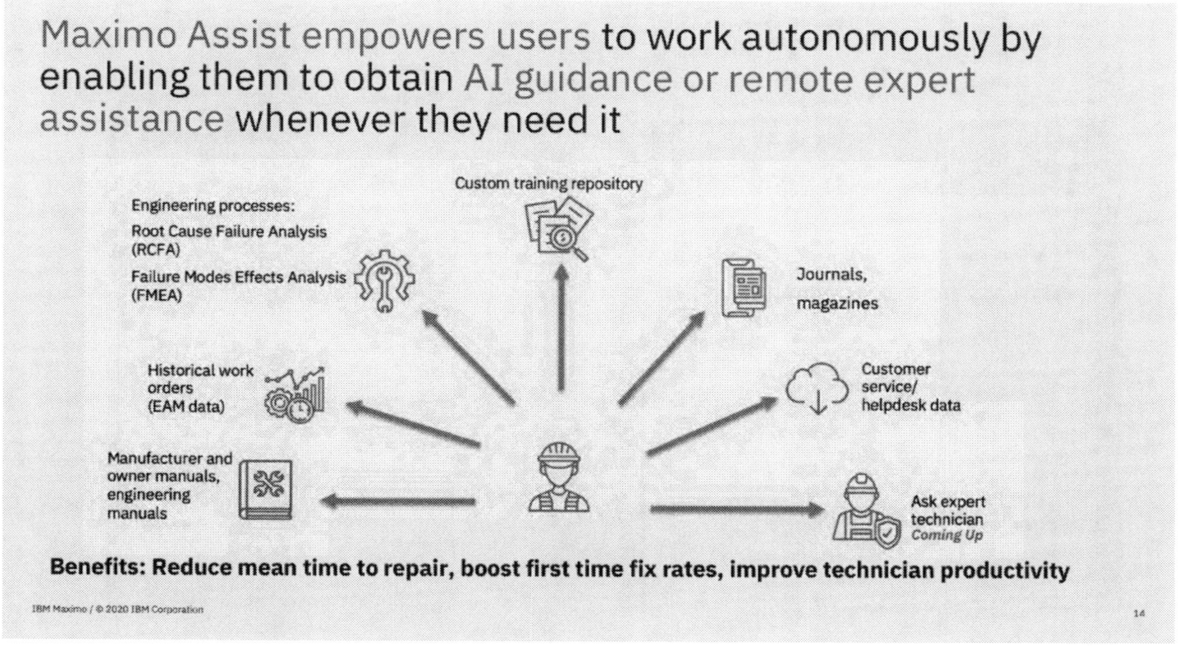
Maximo Assist empowers users to work autonomously by enabling them to obtain AI guidance or remote expert assistance whenever they need it
Custom training repository
Engineering processes:
Root Cause Failure Analysis (RCFA)
Failure Modes Effects Analysis (FMEA)
Journals, magazines
Historical work orders (EAM data)
Customer service/ helpdesk data
Manufacturer and owner manuals, engineering manuals
Ask expert technician
Coming Up
Benefits: Reduce mean time to repair, boost first time fix rates, improve technician productivity
IBM Maximo / © 2020 IBM Corporation
14

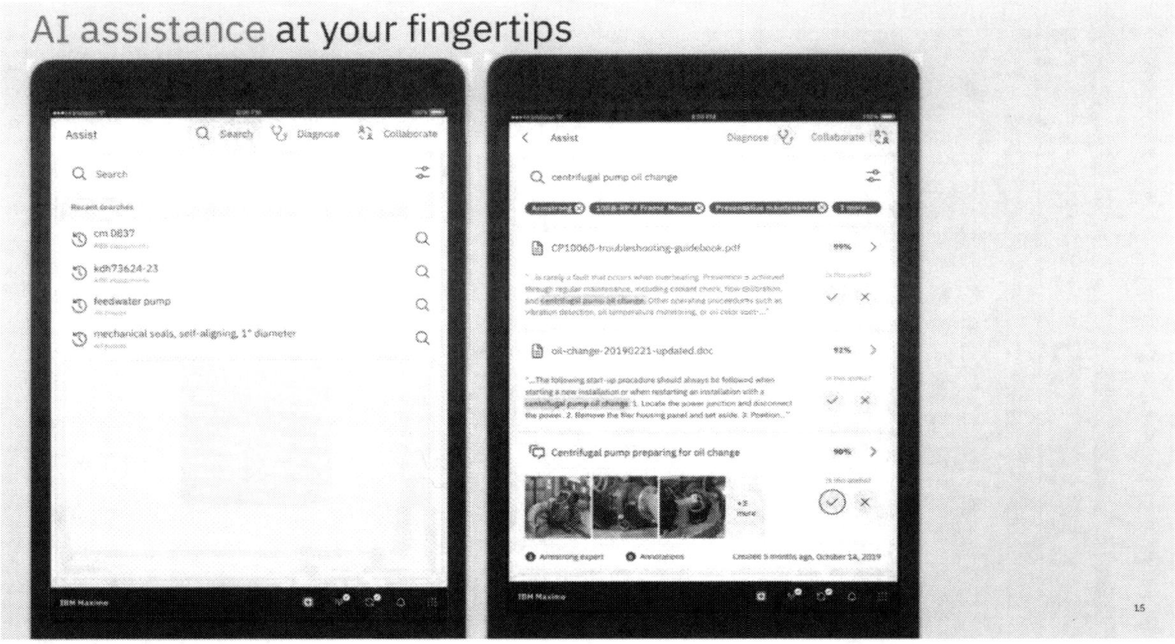
AI assistance at your fingertips
Assist
Search
Diagnose
Collaborate
centrifugal pump oil change
CP10060-troubleshooting-guidebook.pdf
oil-change-20190221-updated.doc
Centrifugal pump preparing for oil change
IBM Maximo
15

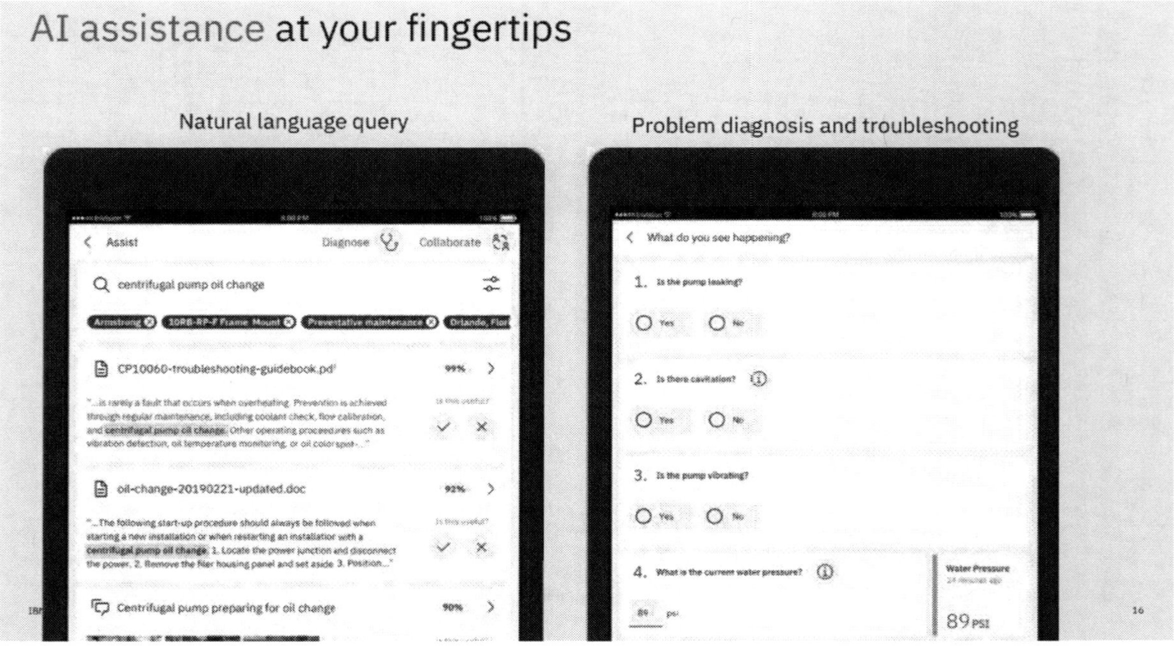
AI assistance at your fingertips
Natural language query
Problem diagnosis and troubleshooting
Assist
Diagnose
Collaborate
centrifugal pump oil change
CP10060-troubleshooting-guidebook.pdf
99%
oil-change-20190221-updated.doc
92%
Centrifugal pump preparing for oil change
90%
What do you see happening?
1. Is the pump leaking?
2. Is there cavitation?
3. Is the pump vibrating?
4. What is the current water pressure?
Water Pressure
89 PSI
16

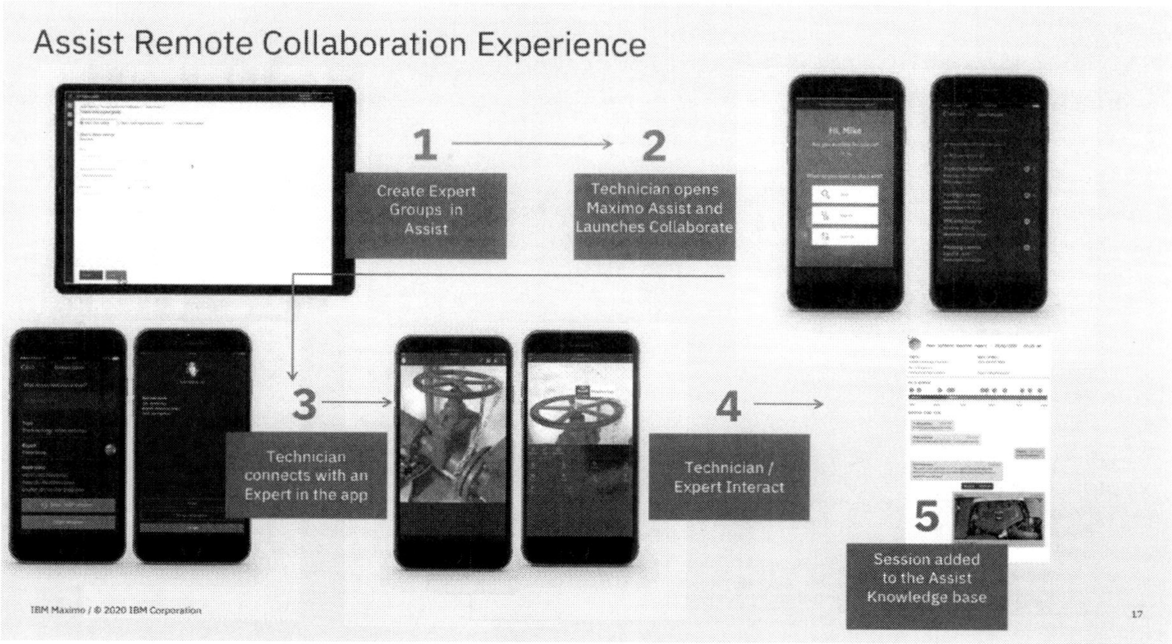
Assist Remote Collaboration Experience
1
2
Create Expert Groups in Assist
Technician opens Maximo Assist and Launches Collaborate
Hi, Mike
3
Technician connects with an Expert in the app
4
Technician / Expert Interact
5
Session added to the Assist Knowledge base
IBM Maximo / © 2020 IBM Corporation
17

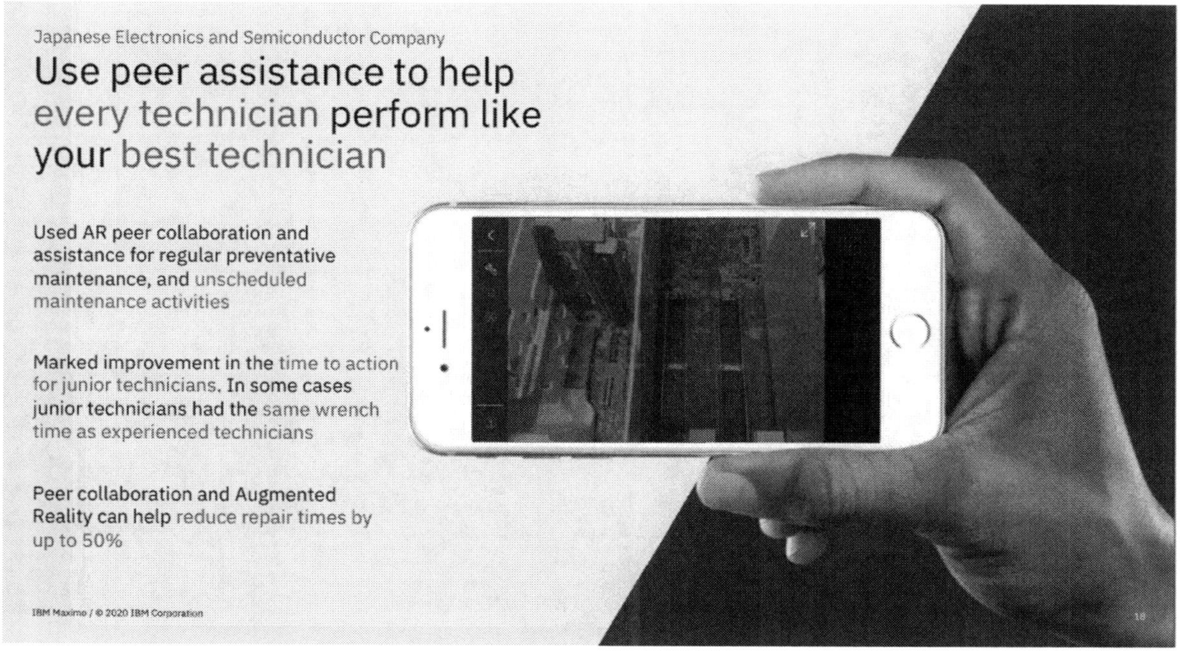
Japanese Electronics and Semiconductor Company
Use peer assistance to help every technician perform like your best technician
Used AR peer collaboration and assistance for regular preventative maintenance, and unscheduled maintenance activities
Marked improvement in the time to action for junior technicians. In some cases junior technicians had the same wrench time as experienced technicians
Peer collaboration and Augmented Reality can help reduce repair times by up to 50%
IBM Maximo / © 2020 IBM Corporation
18

## Empower your experts and scale tribal knowledge seamlessly

Session history saved for future reference and added to AI knowledge base

Capture your tribal knowledge and make it searchable next time a similar incident occurs

Enhance experience for new technicians by giving them tools to do the right fix the first time

IBM Maximo / © 2020 IBM Corporation

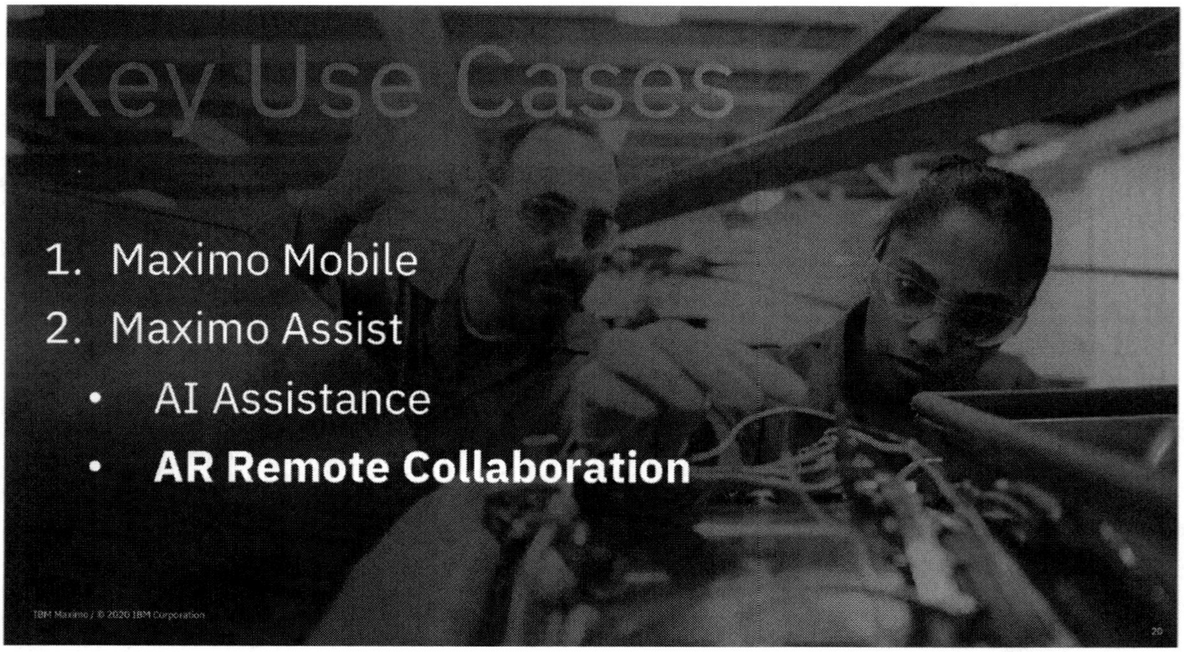

Australia's largest independent oil and gas company

## Aging workforce, 80% of time spent researching problem

Absorbed over 600,000 pages of documentation regarding drilling equipment

Time spent on researching reduced by 75% which equates to $10 million AUD worth of time savings

Client has successfully created a bridge for knowledge transfer from past to present

## Maximo Assist Benefits

- Improve technician productivity
- Boost first time fix rates
- Reduce troubleshooting time
- Reduce Training costs
- Reduce Repair Costs
- Maximize equipment uptime
- Reduced overall maintenance costs

**25%**

Increase in First Time Fix

**$11.2M**

Annual savings in cost avoidance from incorrect procedures and repeat failures

**$10M**

Annual savings from reduced mean time to action

# Connect with us

Heena Purohit
http://linkedin.com/in/heenapurohit/

Kyle Harding
https://www.linkedin.com/in/kyle-harding-00391911/

IBM Maximo / © 2020 IBM Corporation 23

IBM Maximo / © 2020 IBM Corporation 24

Delek US
Refinery of the Future
1

Command Center
The Connected Worker
Operations
Maintenance
GoArc
LOG IN
2

Command Center:
Refinery of the Future
Safe Connected Worker
Aggregation
Delek US
3

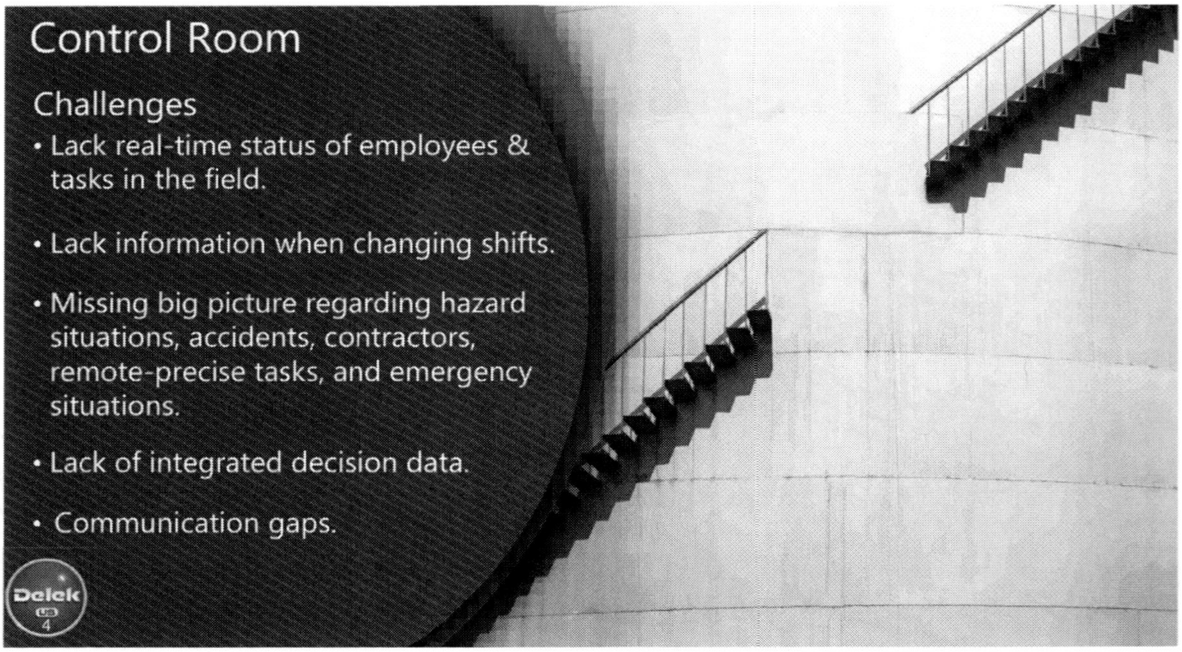
Control Room
Challenges
• Lack real-time status of employees & tasks in the field.
• Lack information when changing shifts.
• Missing big picture regarding hazard situations, accidents, contractors, remote-precise tasks, and emergency situations.
• Lack of integrated decision data.
• Communication gaps.
Delek US
4

Command
Center
Screen
Pending
In Progress
Completed
On Hold
Cancel
5

GOARC Show Me - 127 - 00:35
GOARC
Local
Remote
Real-time
Communication
Command Center
with Operators
6

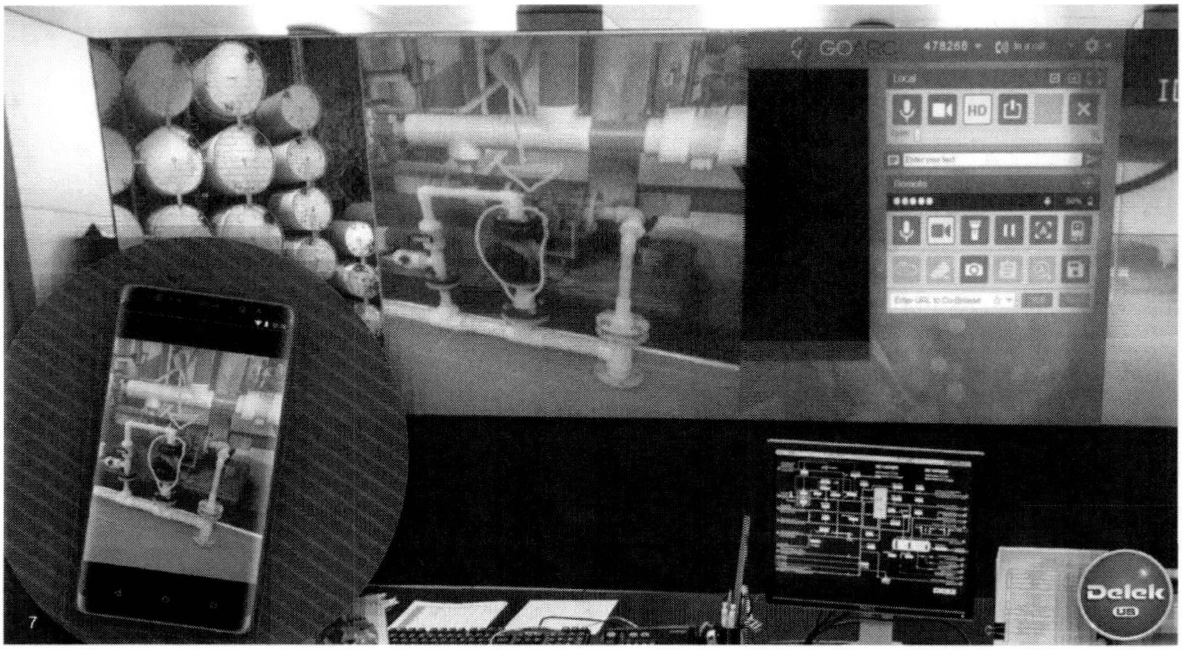
GO
478260
Local
HD
Remote
Delek
US
7

Percepto
Delek
US
8

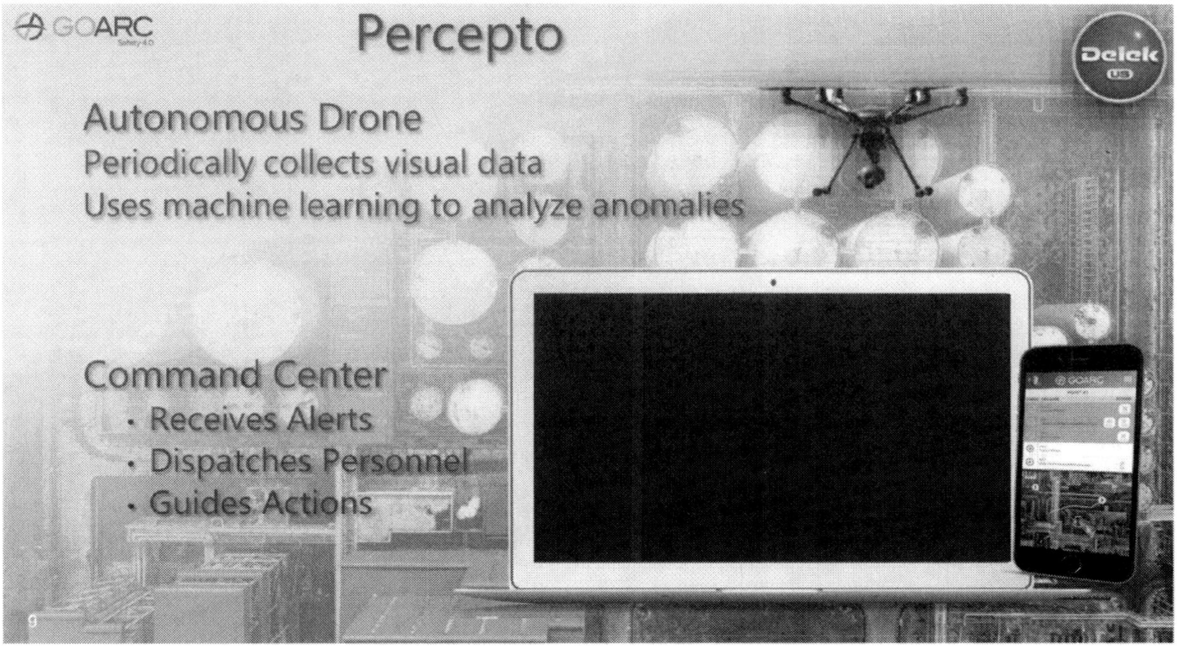
GOARC
Percepto
Autonomous Drone
Periodically collects visual data
Uses machine learning to analyze anomalies
Command Center
• Receives Alerts
• Dispatches Personnel
• Guides Actions
Delek
9

Command Center
Emergency Management Screen
Chemical Spill Accident on Route 4
GOARC
Delek
10

Tracking & Scheduling: progressive measurement percents, levels
Next Desiccant Change:
Scheduled September 14, 2020
Mostly Blue & New
Fading from Blue to Pink
Fading from Pink to Blue
Mostly Pink & Old
Clear & Bright
Dark/varnished
Next Oil Change:
Scheduled July 29, 2020
11
GOARC

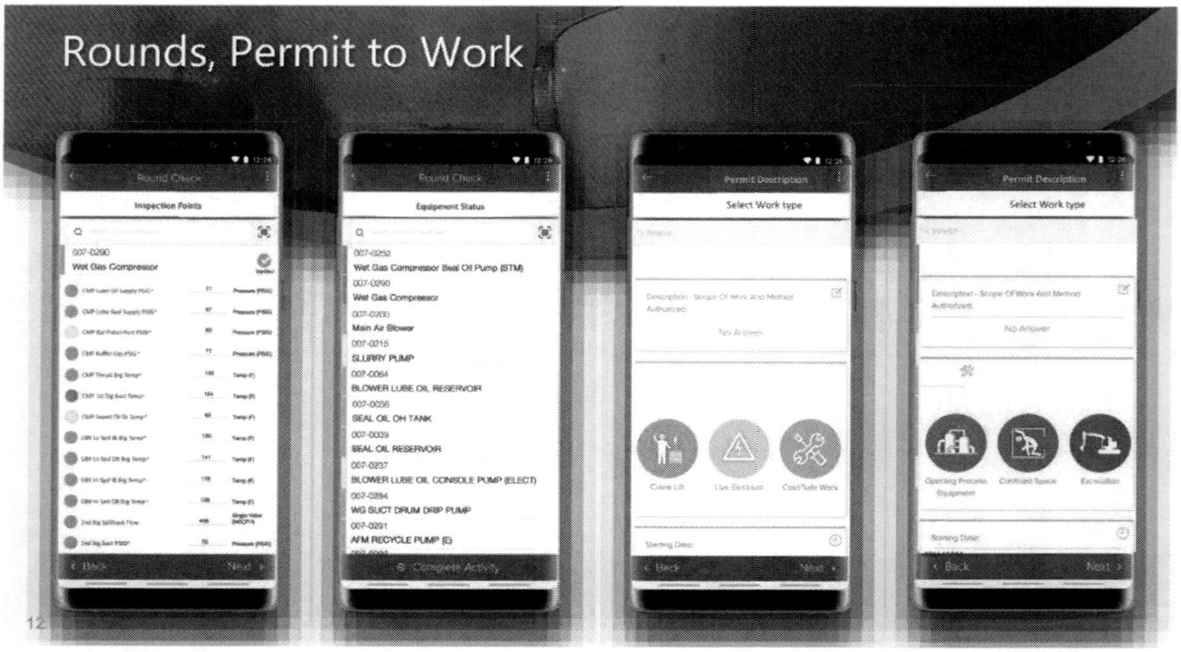
Rounds, Permit to Work
Inspection Points
007-0290
Wet Gas Compressor
Equipment Status
007-0252
Wet Gas Compressor Seal Oil Pump (STM)
007-0290
Wet Gas Compressor
007-0200
Main Air Blower
007-0215
SLURRY PUMP
007-0064
BLOWER LUBE OIL RESERVOIR
007-0056
SEAL OIL OH TANK
007-0039
SEAL OIL RESERVOIR
007-0237
BLOWER LUBE OIL CONSOLE PUMP (ELECT)
007-0284
WG SUCT DRUM DRIP PUMP
007-0231
AFM RECYCLE PUMP (E)
Complete Activity
Select Work type
Select Work type
12

# Management Dashboard

## Reporting

- What is being inspected?
- How frequently?
- What are the issues?
- What is the severity?
- What tasks were completed?
- What measurements missing?

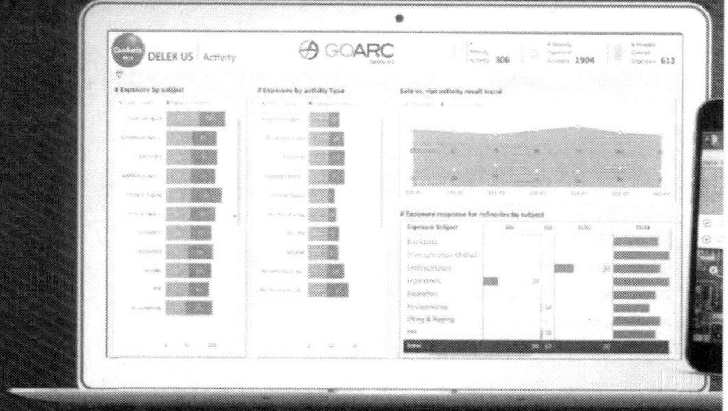


13

# Operations Dashboard

## Reporting

- Measurement guidance
- Leading KPIs
- Diagnostic analytics
- Predictive statistics
- Prescriptive guidance
- Dynamic analysis

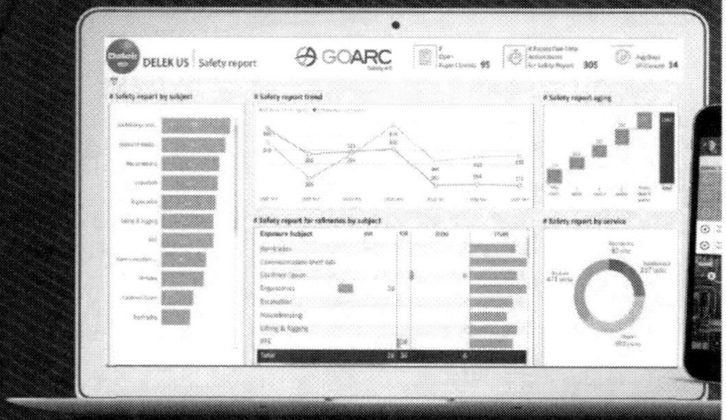


14

1
FRONT LINE OPERATORS
Mobile Application (iOS & Android)
• Paperless Data Collection
• Micro-Learning
• Digital Work Instructions
• Collaboration
2
COMMAND CENTER DECISION MAKING
Command Center Monitor (Web-Based)
• System Integration
• Analytics
• Procedure Authoring
• Job Scheduling
3
C LEVEL REPORTING
Content Management System
• Threat Analysis
• Risk Reporting
• Global Comparisons
15
Delek US

Thank You
Delek US
GOARC
Safety 4.0
16

Plant of
the Future
Virtual Meeting
December 10
Industry Round Table
December 10, 2020
POTENTIAL ENERGY
SELECT LOCK TYPE
Interested?
Contact us:
1. CCPS Exhibit Hall
2. GOARC
3. Chat: name
Delek
17

# AUTHOR INDEX

Abonyi, J. .... 68
Aldeeb, A. .... 40
Biegl, M. .... 124
Buque, C. .... 110
Chappell, J. .... 117
Colegrove, L. .... 11
De Goede, E. .... 87
Dew, P. .... 21
Dorgo, G. .... 68
Downes, T. .... 40
Drenth, D. .... 124
Gimenez, F. .... 29
Groenenboom, B. .... 101
Harding, K. .... 139
Hoedemakers, G. .... 87
Jones, S. .... 60
Krishna, K. .... 50
LaBella, L. .... 151
Lemos, K. .... 1
Nesmith, G. .... 50
Palazoglu, A. .... 68
Pawulski, R. .... 50
Peeters, O. .... 50
Petrusich, J. .... 151
Purohit, H. .... 139
Ranchet, N. .... 95
Ruppert, T. .... 68
Schwarz, H. .... 75
Shahmohammadi, S. .... 87
Van Den Hurk, J. .... 87
Van Middelaar, J. .... 87